The

Great British

Tree

Biography

To Grace and Jackson,
for many more welly walks to come...

Mark Hooper

The

Great British

Tree

Biography

50 legendary trees and the tales behind them

PAVILION

Introduction

Introduction

This is not intended to be a guidebook to trees, famous or otherwise. Nor is it simply a list of our nation's most venerable – and venerated – species. Their botany, their age and their appearance are all secondary here. Which is not to say they are unimportant – rather that there are plenty of exhaustive, definitive books on those subjects already, from Thomas Pakenham's *Meetings With Remarkable Trees* to Will Cohu's *Out of the Woods*. In contrast to such arboreal expertise, I am an enthusiastic amateur, gazing up into the branches in naïve wonder and attempting to untangle the secrets and stories that they hide.

The Great British Tree Biography instead attempts to explore that space where social history meets natural history – and examine how the two are inextricably linked. Our trees have shaped us as people, just as we have shaped them. Britain's forests have formed the backbone of its diverse empires and been the catalyst for some of its most pivotal moments – the oaks that built its navy; the yews, ashes and elms that brought victory to its archers (although, as it transpires, not necessarily at Agincourt); the willows that have produced its cricket bats; even the woodland shortages that forced the switch to coal and fired the Industrial Revolution.

But equally, our trees have been rallying points, under which trysts were furtively pursued, unions were formed, oaths made and sedition planned. Some of those plotters, no doubt, were later hanged from the same boughs. So the tree that might be a symbol of hope to some might bear the heavy silhouette of oppression for others.

Essentially, then, this is a history book: but told through individual trees where something of note once happened – ranging from the sycamore under which the Tolpuddle Martyrs met, to the one that was the site of Marc Bolan's fatal car crash; the oak beside which Wilberforce proposed abolitionism, to the one Paul McCartney jumped into in the video for 'Strawberry Fields Forever'.

Already you can see how this becomes an exercise in separating fact from myth and muddled, half-remembered stories. Many – such as the tale of how a Welsh forest sprung from timber imported from the trenches of Flanders – swiftly petered out into unsubstantiated rumour almost as soon as they had been mentioned. Others – including the story of how a group of copper beech trees in Avebury inspired the Ents in J.R.R. Tolkien's *The Lord of the Rings* – have little in the way of solid proof but are convincing enough to include.

But some myths take hold so firmly they become part of our history, whether we believe them or not. Take the Major Oak in Nottinghamshire's Sherwood Forest. Its inclusion here felt necessary – not just for its location, or for its grandeur (it is so vast the giant, drooping branches have to be propped up with struts) – but for sheer chutzpah. This, we're told, is the very tree that Robin Hood and his Merry Men used as their headquarters. It's certainly old enough to have been around in the reign of King John, and it's nice to imagine Will Scarlet perched on one of its great limbs – even if he never existed. But when other great characters in British history – such as Alfred, Lord Tennyson – add their own layers to the myth, it almost becomes self-fulfilling. (And, talking of cutting through the rings of history – there are trees contemporary to the Major Oak that were found to bear the branding marks of King John, subsumed almost half a metre (18in) under the outer bark over the course of history.)

While we're on the subject of legends, how about the Glastonbury Hawthorn, which flowered on Christmas Day, and was believed to have taken root when Joseph of Arimathea struck his staff upon the ground. (Joseph also brought the Holy Grail to our shores for King Arthur's knights to seek – of course.) And then there's Oswald's Tree: where the dismembered body of Oswald, the Christian King of Northumbria, was said to have been hung by Penda, King of Mercia, as a warning to others – and from where the town of Oswestry in Shropshire takes its name.

Some – such as the aforementioned sycamore next to which the Tolpuddle Martyrs formed one of the country's first trade unions in 1834 – are relatively easy to find. (It, and the small triangle of grass it stands on, represents the National Trust's smallest property.) Others have been faithfully surveyed and recorded, such as those marked 'Tree to Remain' in the plans for the A303 that follows an ancient route from Basingstoke in Hampshire to Honiton in Devon. Yet others, of course, have proven more elusive. Many have fallen victim to disease, storms, fires, the woodsman's axe or vandalism.

The response of the British people to such incidents speaks volumes for the regard with which we hold our trees. The Great Storm of 1987 (when an estimated 15 million trees were lost) was treated as a national trauma, as has the spectre of Dutch elm disease that has killed over 60 million trees in Britain over two epidemics – the first in the 1920s and the second (ongoing) since the 1970s.

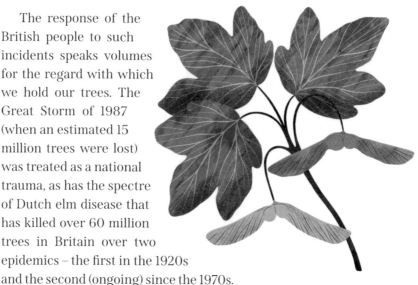

Certainly in the twentieth century, the reaction of the state to such momentous events has been to engage in ambitious projects of extensive tree growing. The Forestry Commission, for instance, was created in 1919 as a direct response to the effects of the First World War, when English woodland had been depleted to an extent not seen since the sixteenth century in order to serve the trenches. Likewise, The Tree Council was initially formed in 1973 to encourage an ongoing campaign of national tree planting in the wake of Dutch elm disease.

As recently as 2011, when David Cameron's minority Conservative government (traditionally trading on the language of patriotism and – ironically, as it turns out – using the symbolism of the oak tree as its party logo) proposed to lease and sell off most of England's 638,000 state-owned forests, the public reaction was as if an act of high treason had been committed. Writing in the *Daily Mail* at the time, Max Hastings, the journalist, author and former editor of the *The Daily Telegraph* and *Evening Standard*, stated, 'It is suicidal for ministers to allow themselves, fairly or unfairly, to appear as foes of the English tree, what the poet Alexander Pope called "a nobler object than a prince in his coronation robes".'

The long, convoluted and often murky history of Britain's relationship with its trees is thanks in part to its peculiar approach towards public and private spaces. Our idiosyncratic systems of hedgerows and our inconsistent history of clearances and Inclosure Acts, have contributed to a romantic, idealized notion of nature in general, and woodland in particular. Our language is full of the metaphor and symbolism of trees. We draw reassurance from the strength of the oak, the immortality of the yew, the fertility of lime and hawthorn and the healing properties of willow.

But often our veneration of trees tips over into anthropomorphism. When the Fallen Oak in Richmond Park was shortlisted for the Woodland Trust's 2019 Tree of the Year competition, it swiftly became a social media meme. The image of a mighty oak, felled by a storm but still clinging to life, its fulsome branches sprouting from its horizontal trunk, was used as shorthand for hope and determination. The British never give up, it seemed to say to us: our doughty resolve embodied in a national emblem. But in reality, nature knows nothing but to grow. Indeed, as John Stewart Collis notes in his post-war classics *While Following the Plough* (1946) and *Down to Earth* (1947), nature will readily strangle itself, given half a chance:

> 'I have come upon portions of the wood where honeysuckle had practically taken over: the captive, the twisted, the mutilated, the dying, the dead ash trees stood hopelessly entangled in the network of ropes, pulleys, nooses, loops, ligatures, lassos which outwardly appeared as lifeless themselves as pieces of cord, but were centrally bursting with life and power, ready and willing to pull down the wood.'

Through his refreshingly unromanticized eyes, Collis sees no need to filter this in terms of cruelty – or through Thomas Hardy's notion of 'Unfulfilled Intention'. It is simply what nature does, regardless (and oblivious) of how that might make us feel...

> 'I came across the same sort of thing every day in my wood. It could make me silent and it could make me sad, but personally I cannot see the spectacle in terms of unfulfilled intention save superficially. What I see is – an almost liquid surging up of life. I see that life as a massive unity, moving and flowering

under the influence of Fire – the air itself taking visible shape in the plants. Some of it does not get up, all of it cannot get up. But if one tree succeeds, one baby survives, I applaud.'

All that said, if we were to apply human characteristics to our trees – if they could observe, if they could understand, if they could speak – what stories would they tell us?

These are the tales that *The Great British Tree Biography* attempts to relate… with the hope that they might reveal something about ourselves as a nation at the same time.

After all, as Charles C. Siefert wrote in his 1938 pamphlet on *The Ethiopian's Contribution to Art* (often attributed to the Jamaican Pan-Africanist thinker Marcus Garvey):

'A race without knowledge of its history is like a tree without its roots…'

Symbolism
& Superstitions

The Symbolism and Superstitions Surrounding Various Species of Trees

The Alder

The alder favours wet and swampy conditions, and is among the first species to colonize poor soil in damp areas. Consequently, its waterlogged woods – or carrs – are often thought to hold magical powers and provide the habitat for mythical creatures including fairies. It is no coincidence that fairies are often depicted in the distinctive green clothing created from the dye of alder flowers. The dye also traditionally provided a natural camouflage for the outfits of Robin Hood's Merry Men – and is still used to this day. Its bark, cones and twigs are also used to produce pigments in distinctive, earthy shades of beige and brown (the introduction of copper and iron creating a darker palette).

When it is cut, the pale wood of the alder transforms into a deep orange hue, giving the impression that it is bleeding. Due to this, many people built superstitions around it: an ancient Irish legend, for instance, decrees that it is unlucky for a traveller's route to lead them through alder woods.

In practical terms, because of the tendency of alder to grow in moist environments, it avoids the damp rot of other wood types. If soaked in water over a long period of time, it becomes rock hard, making it ideal for the construction of bridges over rivers and waterways. This property also meant it was an effective material for shields (alder shields were particularly popular among the Celts).

As a result, the alder is often associated with strength, determination, comfort, peace, protection and royalty – particularly a distinctive feminine strength in facing up to challenges. Hence it is often referred to as 'The Goddess Tree'. It is the tree under which doomed lovers Deirdre and Naoise sheltered in Irish mythology.

As well as its use in dye-making, the bark of the alder was also believed to have medicinal properties, and was used in the treatment of fever and internal bleeding.

The Apple

The apple is, of course, the fruit of temptation in the Adam and Eve creation story of the Old Testament. Quite why this otherwise harmless tree, with its abundant crop of sweet, nourishing fruit, should be seen as the symbol of man's corruption is a moot point. Perhaps it was simply the best tasting – and therefore most tempting – food available in the garden.

Otherwise, the apple is associated with fertility and health – its bountiful, easily accessible crop a ready source of refreshment since Roman times, where it was cultivated and so highly valued that there were severe punishments for anyone who cut down an apple tree.

In both Norse and Celtic, as well as Greek mythology, the apple represented eternal youth, while it is often used as a metaphor for sensual love (which perhaps points at the deeper symbolism behind the apple that the serpent enticed Eve to bite).

Most obviously, trees heaving with green and red apples are a sure sign of the arrival of the harvest. Hence the pagan tradition of wassailing, which is still celebrated in the cider orchards of Somerset during January. Here, the coming harvest is blessed at dusk in a ritual that involves dousing the lower branches and roots of the oldest apple tree with cider. Pieces of toast are also hung in the branches to attract robins (thought to bring luck and protection). The assembled crowd then bows three times, slowly rising with exaggerated effort, as if weighed down by heavy loads. Finally, pots and pans are noisily clattered together and shotguns are blasted into the branches to frighten away evil spirits and to awaken the tree from its winter slumber. Songs and chants hail the apple tree before its produce – richly fermented – is sampled with ready abandon.

The Ash

The ash tree has long been thought to have mystical healing properties and the wood was burned to ward off evil spirits. It was also commonly regarded as an antidote to various poisons, including the bite of venomous snakes. Numerous superstitious rituals involve passing living mammals – from rodents to human children – through the split branches of an ash tree (or even buried within the trunk, in the case of field mice) in order to cure lameness, cramp, rickets and 'rupture' in both children and cattle. The sap from a burnt stick of ash was once traditionally used in Scotland as the first spoonful of food for a newborn baby, intended to protect it from disease.

The extent of the connection drawn between ash trees and the prevention of illness was such that they would be jealously guarded – the death of an ash tree was often thought to imply the loss of protection that it held over its 'patients'.

In Norse mythology, the ash was the 'Tree of Life', from which the first man on Earth emerged. This is echoed in Teutonic legends, where the gods held court under the shade of an ash; its crown ascending to heaven and its roots reaching into hell. To this day, it is sometimes referred to as the 'Venus of the woods'.

The Druids regarded the ash as sacred – a link made overt in Robert Graves' exhaustive but academically flawed book *The White Goddess* (see p.143). The wands used by Druids – a forerunner of wizard (and Harry Potter) legends – were often made of ash due to its straight grain.

The Beech

Where the oak is often considered to be the 'king' of British trees – with all the patriarchal connotations that implies – the beech is associated with femininity and is often presented as the 'mother of the woods' and queen to the oak in Celtic belief. It is often used to represent Diana, who in both Greek and Latin mythology is the goddess of forest and woodlands, hunters, and the moon, and is associated with protection.

The beech also signifies prosperity and wisdom – a result of the ancient practice of using thin slices of beech as an early form of papyrus. (The Anglo-Saxon word for beech – *bok* – is thought to form part of the etymology of the modern English 'book'.)

The tree was traditionally believed to have medicinal properties; its leaves boiled to make a poultice that was used to relieve swellings and various skin conditions. It is a natural antiseptic – obtained by the dry distillation of the branches into a tar or creosote. (This tar has also been used to treat the pain of toothache.)

Because the smooth bark is easy to carve, beeches are often used as lovers' trysting trees – with initials, hearts and other symbols of affection drawn onto the trunk. (See pp.91–93.)

The smooth bark also explains the superstition that a beech tree can never be struck by lightning. Of course they are just as likely to be struck as any other tree, but they rarely show signs of damage as the bark – together with the fatty content of the wood and the fine branches – make it an extremely good conductor of electricity, channelling the lightning to Earth.

The Birch

While it may have several positive connotations – its leaves contain vitamin C, for instance, and it is used in the treatment of urinary and kidney infections – the birch is forever sullied by its connection to witchcraft. The reason for this is innocent enough – traditional household 'besom' brooms (which would be found in all households) would customarily have bristles made of a bundle of birch twigs (usually attached to a handle of either hazel or hawthorn).

Besom brooms were, of course, also the broomsticks used by witches. As well as the legends of witches flying on their broomsticks, besoms are also used in Wiccan ceremonies, sweeping out a space to purify it and remove negative energies before a ritual ceremony.

You can see the problem here. As with many of the supposed signs of witchcraft, there is little to differentiate an everyday household object from a symbol of the occult. Consequently it was often at the whim of the accuser to decide whether or not someone was a witch.

Ironically, the birch is also believed to ward off evil (cradles were traditionally made from birch to protect children) and is another species known as the Goddess Tree, with its connections to Venus, the White Goddess, Freya and Brigid. It is associated with beauty and tolerance – again, it becomes obvious that, threatened by an all-powerful feminine deity, the defenders of a patriarchal society might invert the properties associated with the birch: thus a symbol of purity and goodness is posited as evil; beauty becomes ugliness; protection becomes threat.

The Elm

Elm trees are often associated with melancholy and death, traditionally being the preferred choice for coffins and often planted around burial mounds and believed to guard the passageway to the afterlife. In Greek mythology, the first elm tree is said to have sprung at the spot where Orpheus played a love song on his harp to celebrate rescuing Eurydice from the Underworld.

This paradox of grief and joy surrounding death is encapsulated in the elm: it symbolizes wisdom and the continuous cycle of life, death and rebirth – and consequently is often used in memorial plantings, particularly those commemorating wars.

The connotations with death probably stem from the fact that elm trees can drop their dead branches without warning – and also the characteristic pliability of the wood, meaning it can bend and distort easily (as if the branches are bowing out of respect). Indeed the etymology of the 'wych' in wych is from the Old English *wice*, meaning 'supple' (and also gives us the derivatives 'weak' and 'wicker'). Understandably, the wych elm is often wrongly thought to be connected to witches (such as Bella's Wych Elm – see pp.78–80). In fact, in all myths witches are deterred from coming near elms due to their role of offering protection from dark forces (elves are also said to live in elm woods).

Due to its pliability, elm is poor at supporting loads, but has other practical uses – not least in shipbuilding, as, like alder, it tends not to rot or split in water. It can also be bent to form shelters and was the popular choice for bow-making among Welsh archers.

The elm's associations with death, of course, are etched into the modern British psyche too: the devastating effects of Dutch elm disease, caused by a fungus spread by elm bark beetles, has left its mark on the landscape. Following an outbreak first identified in 1927 in England, a second, far more virulent strain of the disease is thought to have arrived via shipments of diseased elms from Canada – probably first occurring in 1967, but not formally identified until 1973. This strain spread rapidly, reaching Scotland from the first case in Hampshire within a decade, and is still prevalent today (if now managed). In total, 25 million elm trees are believed to have been killed, effectively wiping out a once common feature of the British countryside (as prominently featured in Constable's *Salisbury Cathedral from Lower Marsh Close*).

The Hawthorn

The hawthorn, with its abundant blossom of white flowers appearing in May, has long been associated with the celebration of spring. It is a pagan symbol of fertility and has an ancient connection with May Day. It features prominently in Beltane, the ancient Gaelic festival held on 1 May; hawthorns were the original maypoles around which dances are customarily performed to mark spring's arrival (with its blossom still traditionally adorning the top of the maypole). The leaves and flowers of the hawthorn are used in May Day garlands and also make up the wreath of the Green Man – the pagan symbol of nature's rebirth, which developed in parallel in a number of ancient cultures. The hawthorn was a symbol of hope to the ancient Greeks, who hung garlands of its white blossom over their doorways during their own May Day festivities.

As a symbol of rebirth, the ancient legends of the hawthorn have inevitably become intertwined with the story of the Resurrection in the Christian faith. The Crown of Thorns that Christ wore on the cross is sometimes depicted as hawthorn (indeed, the spiny hawthorn is native to Israel), and there are superstitions that tell of the hawthorn making sounds of groaning and sighing on Good Friday.

The hawthorn is also at the heart of the myth of the Glastonbury Holy Thorn (see pp.70–71), in which Joseph of Arimathea landed in Avalon (an island located on the modern-day Somerset Levels) and preached Christianity to ancient England. The story goes that he planted his hawthorn staff into the ground where it is said to have taken root – the progenitor of the Holy Thorn that flowers every Christmas Day.

This is not, however, the derivation of the name 'mass tree' or 'mass bush', as the hawthorn is still known in areas of Ireland. Rather, this refers to the rites of Mass in the Catholic Church and is associated with the saints, burial rituals and the folklore of the 'little people'. (Throughout the United Kingdom, myths tell how fairies live under hawthorn bushes.)

Paradoxically, despite all its connotations involving life and rebirth, the hawthorn has also forever been associated with disease. While it may have garlanded doorways since ancient times, hawthorn was never brought inside the home, due to the superstition that it brought with it illness and death. In medieval times it was said that hawthorn blossom carried with it the smell of the Great Plague. This has been explained by modern botanists, who identified the chemical trimethylamine in hawthorn blossom, which also forms in decaying animal tissue: thus the hawthorn does literally smell of death.

The Lime

The lime tree (or linden) has associations with fertility in many cultures. Due to its distinctive, heart-shaped leaves, the lime is often a symbol of love, fidelity and justice.

Despite these links with setting the heart aflutter, the dried flowers of the lime are actually a natural sedative, used in calming nerves, lowering blood pressure, relaxing spasms and improving digestion. Lime-flower tea is still used as a cure for colds and for treating hypertension, digestive and cardiovascular complaints. This perhaps explains why cattle enjoy eating the green leaves of lime trees. In rather more prosaic terms of fertility, the lime can also play an important role in agriculture, as it is a nitrogen-fixing tree, improving soil structure.

In France and Switzerland, limes are a symbol of liberty and

are often planted to commemorate battles. In various central European mythologies, including pre-Christian Germanic, Slavic and Baltic, the lime or linden is a sacred tree. In Germany, the linden was a symbol of justice, with meetings of a legal nature undertaken beneath the cover of its branches.

The Mulberry

The Mulberry is found commonly throughout Europe, Asia and the Middle East. Consequently, numerous myths and legends about it have sprung up around the world. It is known as a symbol of faith, growth, death and nature itself.

The mulberry has remarkable medicinal properties. Its leaves have been used to treat illness and disease for centuries. A Victorian medical journal lists that its berries 'are good to stay fluxes... and a syrup of the berries cures inflammation of sores of the mouth, throat and palate, when it is fallen down.' The bark of the root is said to expel tapeworm from the gut, whilst 'a decoction made of the bark and leaves is good to wash the mouth and teeth when they ache. The leaves of mulberries applied are said to stay bleeding at the mouth or nose, or of the piles, or of a wound.'

The tendency for mulberries to wait until all frost is gone before coming into bud led the ancient Greeks to dedicate the tree to Athena, goddess of wisdom. In Chinese symbolism, the mulberry is a sacred tree linking the Earth to heaven, upon which the sun bird perches. Paper made from mulberry is still used in Shinto religious ceremonies in Japan, while its leaves are the favoured food of silk worms. This prompted the famous failed attempt by King James I to instigate an English silk industry by importing mulberry trees to these shores in the early seventeenth century (see p.149). Sadly, the wrong variety was brought to these shores and the silkworms refused to eat the leaves.

In fact, the mulberry was first introduced to Britain over 1,500 years earlier: it is thought that the Romans actually brought the tree with them during their Britannica campaigns of AD 43–84 – its various healing properties important for an army that marched on its (healthy) stomachs as well as its feet.

The Oak

The mighty oak, indigenous to Great Britain and Ireland, has long been used as a symbol of the nation itself, particularly drawing on the metaphor of its strength and obduracy – symbolic of an unbowing unity of purpose.

The mythology surrounding the oak is associated with numerous British tribes and eras – the Celts, the Druids, the Norse, the Romans, the Anglo-Saxons and the Normans all worshipped or honoured the oak in their own ways. In ancient Britain, the Druids practiced specific rituals in oak groves, and lent particular significance to the mistletoe that grows in the branches of a host oak tree – using it to cure infertility as well as an antidote to poison.

For the Viking invaders from the Norse kingdoms, oak was revered as the sacred Thunder Tree, protecting those sheltering under it from lightning sent by the god Thor. It held a similar significance in Greek mythology, where it was the sacred tree of Zeus, king of the gods on Mount Olympus and the god of sky and thunder. The Slavic equivalent, Perun (god of thunder, storms, law and fertility), was also symbolized by an oak.

As well as gods, the oak is also linked with royalty, with ancient kings and Roman emperors wearing crowns of oak leaves (up to the present day, an oak leaf is used as a symbol of merit for the armies of numerous countries, including the USA, Germany and members of the Commonwealth).

Early Christianity drew on the pagan tradition that the oak provided protection from evil and warded off the effects of witchcraft. One example tells that King Ethelbert of Kent, who ruled AD 550–616, advised St Augustine, a founder of the English church who became the first Archbishop of Canterbury, to always preach under an oak tree. A tree known as Augustine's Oak (now replaced by a stone cross) once stood at the point in Thanet where King Ethelbert first met the saint. An inscription at the foot of the stone (in Latin and English) reads:

'After many dangers and difficulties by land and by sea, Augustine landed at last on the shores of Richborough in the Isle of Thanet. On this spot he met King Ethelbert and preached his first sermon to our countrymen. Thus he happily planted the Christian faith which spread with marvelous speed throughout the whole of England – AD 596'

As a representation of knowledge, the mystical 'Tree of Life' is often portrayed as an oak (or otherwise an ash) – its roots penetrating into the Underworld while its branches reach into the heavens. From ancient times, the tree was also believed to possess restorative and curative powers, with ailments treated variously by rubbing the trunk with one's palm or driving a nail into it. Historically its bark was used in the treatment of sore throats, obstructions of the liver and 'alleviating the spasms of gravel' (gallstones), while the water that gathers in oak tree hollows was used to alleviate sores, scabs, rashes and lesions.

The tradition of kissing under the mistletoe is linked to the oak that often hosted it. Couples followed the custom of marrying under the boughs of an ancient oak (the 'marriage oak') until it was forbidden by the Christian church – however, it remained a common tradition that the newlyweds would dance three times around an oak tree after their marriage. (And a drink made from an 'infusion of crushed acorns' would sometimes also be drunk.)

The distinctive rounded leaves of the oak, meanwhile, have been used as the emblem of many conservationist and environmental groups, including the Woodland Trust and the National Trust.

The Sycamore

The ancient Egyptians used the sycamore to symbolize the connection between the worlds of the dead and the living. In *The Egyptian Book of the Dead*, sycamores offer 'blissful repose' to departed souls.

Oddly, despite its great age, and associations with the Holy Land, few myths have sprung up around the sycamore since its arrival in Europe. It is a symbol of perseverance, vitality and longevity – but these are all obvious attributes of the tree itself.

The sycamore's sturdy, strong wood made it popular for furniture-making, but also lent itself to a more sinister use: sycamores were often used as hanging or dule trees (see pp.100–102). Consequently, there are frequent occurrences of the sycamore in poems, songs, laments and folklore surrounding death. The words to 'The Willow Song' (c.1583), as quoted by Desdemona in Shakespeare's *Othello*, begins 'The poor soul sat sighing by a sycamore tree...'

Poor souls have been depicted sighing under sycamores for centuries – although when exactly the tree was introduced to Britain is a matter of some debate. One theory suggests that knights returning from the Crusades in Palestine brought sycamore seeds with them, where the tree was considered sacred.

The Willow

Although today the willow tree is associated with ideas of sorrow and mourning, in ancient times willows were regarded as trees of celebration, and willow branches form part of the ritual for the Jewish festival of Sukkot or the Feast of Tabernacles.

Looking further back, Druidic lore tells that the Universe was hatched from two snake's eggs hidden in a willow tree. In Celtish, Druidic and Greek mythology, the willow is used in prophecy. The Greeks also believed willow bestowed divine inspiration, particularly to artists and poets – as well as being sacred to the Underworld goddesses Circe, Hera, Hecate and Persephone.

This apparent paradox – of being instrumental in the creation of all life whilst also symbolizing death – positions the willow as central to the philosophies and mythologies of countless cultures. In overseeing life and death, it represents immortality and the cycle of nature. The ability of the branches from fallen willow trees to take root where they land is a physical manifestation of this, while its ability to endure enormous pressures, bending against the elements without breaking, provides a metaphor for determination and the endurance of spirit.

It is important not to conflate the many species of willow – almost 400 in all. In common imagination, the weeping willow is most often called to mind – its branches drooping as if in sorrow over their watery habitats. This is the willow that most often appears in poetry and art – witness Shakespeare's *Hamlet*, with Ophelia drowning near a willow tree.

The willow's proximity to water, perhaps, explains why it was traditionally used as a cure for rheumatism and other illnesses associated with damp (perhaps an unconscious understanding that, as with vaccines, a small dose of the source of the illness can elicit the cure). There is, however, medical science to back up the folk medicine: salicin, an extract from willow bark, is used to treat rheumatism today. The painkiller aspirin is also derived from salicin – confirming the wisdom of the medieval practice of chewing willow bark to ease the pain of toothache and headaches.

The bark was also boiled in water, the liquor drunk to relieve diarrhoea, help reduce joint inflammation in arthritis and as a gargle for sore throats. It was also used to stop bleeding, clean wounds and to treat general aches and pains.

The willow, then, is a silent sentinel of wisdom; a mourner and reliever of pain; a symbol of eternity and strength – of life, the universe and everything.

The Yew

The yew is a symbol of immortality and fertility, but is also seen as an omen of death. Like the willow, this apparent duality is perhaps best seen as part of the same whole: an endless, complete circle of life.

Let's talk of death first. Yew trees are often found surrounding churchyards – and yet its association with funeral rites go back further than Christianity. (In fact, there are hundreds of examples of churchyards in England with clusters of yews that predate the churches themselves.)

The popular explanation is that the yew offered protection from the devil, protecting those buried within the circle of distinctive trees. Furthermore, yews were often planted directly above the graves of plague victims to protect and purify their bodies. And yet there may be a far less spiritual explanation for their presence in and around graveyards. The yew's needles are highly poisonous to cattle and other grazing livestock and can be fatal if eaten: by surrounding their places of burial with yew, the clergy may simply have been attempting to stop people from grazing their cattle on church land and defiling consecrated ground.

Shakespeare was clearly aware of the poisonous nature of the yew: in *Macbeth*, the king brews a poisonous potion of 'slips of yew, silvered in the moon's eclipse'. A peculiar legend involving yews in Scotland says that if someone holds a cutting of yew from church grounds in their left hand, they can threaten an enemy without being heard, even though all those around them will hear the words perfectly. Thus they can justly claim to have given their enemy fair warning, without the enemy having been aware of it.

Due to its toxicity, the medicinal uses of yew are limited – although it was famously the wood of choice for Scottish and English bowmakers. (Despite this, the English archers of Agincourt actually favoured Spanish and Portuguese yew for their longbows over the knotty, homegrown variety – see pp.94–95.)

An A–Z
of British Trees

1

Addison's Oak

Oak, Bristol, England

Standing proud in the middle of the Sea Mills estate in Bristol, Addison's Oak was planted in 1919 in honour of Dr Christopher Addison, 1st Viscount Addison, the Liberal MP for Hoxton, Shoreditch and Minister of Munitions in David Lloyd George's coalition government during the First World War. In 1917 he was appointed Minister Without Portfolio, charged with the planning for post-war reconstruction. Central to this was Addison's Housing and Town Planning Act, which oversaw the building of the UK's first low-rent council houses for the working classes.

An eminent doctor, he also served as Minister of Health and had identified that poor housing was a major contributing factor to the health problems afflicting ordinary members of the public. Addison's solution was for the state to work with council planners to replace slum dwellings with more adequate social housing – putting an emphasis on the beneficial effects of fresh air, good light and greenery.

Inspired to act after witnessing the squalor of his own constituency in East London, Addison engaged in a nationwide campaign. On 4 June 1919 he cut the first sod of earth for a public housing scheme in Bristol, under the slogan of 'Homes Fit for Heroes', arguing that the country owed those returning from war a better standard of living. Speaking to the crowd, he said they 'did not want houses built in dismal streets' and that 'until they [the British] have houses with air about them, so long would they have to spend enormous sums annually on sickness...' To commemorate the occasion, an oak was planted in his name.

In 1921, Addison resigned when his housing programme fell victim to austerity cuts. Of the 500,000 new homes promised by Lloyd George, only 213,000 were built. The next year, Addison wrote a book on the subject, *The Betrayal of the Slums* – the title deliberately ambiguous, referring both to the original betrayal of the people forced into slum living as well as the subsequent failure of the government to see the housing project through.

However, by 1939, thanks to the pioneering blueprint of the Addison Act, 1.1 million new council homes had been built across the country. Addison himself went on to join the Labour Party, serving as Leader in the House of Lords under Attlee's government of 1945, and assisting Nye Bevan, then Minister of Health, in pushing through more social reforms.

In 2019, on the centenary of its planting, Addison's Oak was shortlisted in the Woodland Trust's Tree of the Year competition, and a special ceremony was held in Addison's honour under the tree that bears his name. As a reminder of his basic truism that investment in decent, affordable housing is also an investment in a happier society, the following quotation was read, written by former poet laureate John Betjeman who visited Bristol's Sea Mills estate in 1937 and remarked upon the:

> 'surprising beauty showing off in the evening sunlight; and vistas of trees and fields and pleasant cottages that that magic estate has managed to create...'

2

The Allerton Oak

Oak, Calderstones Park, Allerton,
Liverpool, England

Believed to have been the site of a medieval 'hundred' court (which would meet twelve times a year), the Allerton Oak in Liverpool is aged at around 1,000 years old. (So there is a chance it was already growing at the time Allerton is mentioned in the Domesday Book in 1086.)

But the distinctive scar that nearly splits the tree in two is the result of a later incident in local history. On 15 January 1864, the *Lottie Sleigh*, a 220-tonne barque (or three-masted sailing ship) was at anchor on the River Mersey off Monk's Ferry near Birkenhead. The ship, built on Prince Edward Island, Canada, was bound for the West Coast of Africa with a general cargo – including 940 quarter-kegs (around 11 tonnes) of gunpowder.

Shortly after 6pm, while he was trimming the wick of a paraffin lamp, the ship's steward accidentally ignited a can of oil, which he dropped on the cabin floor, setting the curtains and bedclothes on fire. The flames spread quickly and, recognizing the gravity of the situation, the steward immediately raised the alarm. Thankfully, a Rock Ferry steamer named the *Wasp* was passing up the river at the time and, alerted by the shouts, its captain Joseph Hughes came up alongside the stricken ship. Thus all hands on board were able to swiftly abandon the *Lottie Sleigh*, with all accounted for (although a second steamer from Rock Ferry, the *Nymph*, reported hearing the barking of a lone dog which had been left on board).

At approximately 7.20pm, the flames finally reached the gunpowder and a huge explosion rent the air. An eyewitness account in the *Liverpool Mercury* the next day records how 'The contents of the vessel blew up with a report hardly possible to describe... Its effects in every part of Liverpool were severely felt and created indescribable terror.' The explosion was heard over 48km (30 miles) away, and the shockwave left a trail of destruction across Liverpool, as the *Mercury* report continues:

'At the same time the most solid blocks of warehouses, offices and private dwellings were shaken to their base – doors locked and bolted were thrown wide open – hundreds, yea even thousands of squares of glass were smashed.' Despite the severe damage and the number of bystanders stood on the shoreline watching the burning ship at the time, miraculously there was no loss of life.

Even though it was some 5km (3 miles) from the *Lottie Sleigh*, the magnificent Allerton Oak in what is now Calderstones Park was also hit by the shockwave from the explosion, causing a huge crack to appear in its trunk. Various attempts have been made to support and strengthen the weakened tree, dating back to 1907 when the first metal props were installed. Today, thanks to an £80,000 investment from Liverpool City Council and The Mersey Forest, a more sophisticated mechanism of adjustable supports has been developed and the tree continues to grow.

As a local landmark, leaves and acorns from the Allerton Oak were sent to soldiers from Liverpool by relatives and loved ones during the Second World War as tokens of good luck. Because of this, many seedlings of the original tree are believed to be growing around the world – not to mention Allerton Oak the Younger, which was planted in Calderstones Park in 2007 from one of the original's acorns. In 2019 the Allerton Oak was voted England's Tree of the Year in the annual Woodland Trust competition.

3

The Ankerwycke Yew

Yew, Runnymede, Wraysbury, Berkshire, England

You will find that a certain pattern emerges in this book in relation to yew trees. Because of their potential to live to a great age (in fact greater than any other plant in Europe), it's almost inevitable that they should become intertwined with local legends.

Even with the difficulty in dating them accurately (since the boughs of ancient yews are often hollow, making a ring count impossible), there are usually contemporary reports that can help us make an educated guess. The fact that yews often predate any nearby dwelling (and can often be the reason any such dwelling is sited there) means they are naturally bound to local history and folklore.

Having established all that, when I tell you that the Ankerwycke Yew sits on an island in the River Thames opposite Runnymede, you'll know what's coming next: legend insists that King John signed the Magna Carta under its boughs. As the one landmark that has survived since 1215, it's inevitable that people try and tie the two together. And who's to say King John didn't admire the grand old tree – which could well have been 1,000 years old even then – as he dipped his seal in the wax on completion of England's first constitution?

But it doesn't end there. We're also told that King Henry VIII courted Anne Boleyn beneath the tree's branches. If that sounds a little more far-fetched, surely it's not unreasonable to think that a man in love would find it a romantic spot, and a fitting place for the King of England to impress his future queen, given its majestic history, whether or not the myth happens to be true.

4

The Birnam Oak

Oak, Birnam Wood, Perthshire, Scotland

[Thunder. Third Apparition: a child crowned,
  with a tree in his hand]
THIRD APPARITION:
'Be lion-mettled, proud, and take no care
Who chafes, who frets, or where conspirers are:
Macbeth shall never vanquish'd be until
Great Birnam Wood to high Dunsinane Hill
Shall come against him.

MACBETH:
That will never be.
Who can impress the forest, bid the tree
Unfix his earthbound root? Sweet bodements! Good!
Rebellious dead, rise never till the wood
Of Birnam rise, and our high-placed Macbeth
Shall live the lease of nature, pay his breath
To time and mortal custom.'
Macbeth, Act IV, Scene I, William Shakespeare

The Birnam Wood prophecy presages a pivotal scene in Shakespeare's 'Scottish play' as Macbeth witnesses the combined armies of Malcolm, Macduff and Siward advancing towards Dunsinane Hill, camouflaged with limbs from the trees. His sense of invincibility drained, he rounds on the messenger who reports:

'As I did stand upon the hill
I look'd upon Birnam, and anon, methought
The wood began to move.'

Macbeth replies:

> 'If thou speak'st false,
> Upon the next tree shall thou hang alive
> Till famine cling thee: if thy speech be sooth,
> I care not if thou dost for me as much.'

It is recorded that Shakespeare did indeed visit Birnam, along with Perth and Aberdeen, during a tour of Scotland in 1599. This was in response to a request sent from King James VI of Scotland to Elizabeth I for 'entertainers' from England.

The great forest that Shakespeare witnessed no longer exists, with the A9 and B867 cutting a swathe through what was once its middle as they follow the River Tay. But on the outskirts of the town stand two of the 100 listed Scottish Heritage Trees. The Birnam Oak and the nearby Birnam Sycamore are the last remnants of the original magnificent woodland that inspired Shakespeare. The sycamore is dated at around 300 years old, meaning it appeared over 100 years after Shakespeare's death, but the sessile oak is thought to date back 500 years, meaning it would already have been impressive in size in the bard's day.

Today, the Birnam Oak bears all the hallmarks of an ancient tree: its huge trunk – with a girth of 5.5m (18ft) – hollow for the first 3m (10ft) above the ground. The largest of its lower limbs has long been propped up by stakes and in recent years it has been in danger of collapse due to the pressure of its heavy branches on the weakened base. But to date – unlike its neighbours witnessed by Macbeth's messenger – it remains defiantly unmoved.

5

The Bolan Tree

Sycamore, Barnes, London, England

Trees often stand as memorials to great figures. Sometimes they are planted in honour of dead heroes, with myth intertwining the truth of what went before. But sometimes a tree becomes an impromptu memorial, not as a way of remembering someone who has died, but because it was the site of their actual death. Such a tree can be found on Queens Ride, Barnes, in London, where an unprepossessing sycamore marks the spot where T. Rex singer/songwriter Marc Bolan died on 16 September 1977.

Bolan was returning from a night out at Morton's, a nightclub and restaurant in London's Mayfair, with his girlfriend, the American singer Gloria Jones (who had a hit with the original version of 'Tainted Love', later covered by Soft Cell). They were less than a mile from Bolan's East Sheen home when Jones lost control of her Mini on a humpback bridge and crashed first into a fence post, coming to rest against the sycamore. Bolan, who wasn't wearing a seatbelt, was killed instantly.

It's hard to overstate how big a star Bolan was in Britain in the early 1970s. With his band T. Rex, he helped to inspire the glam rock movement, with an incredible run of ten consecutive top five singles between 1970 and 1973, including four number ones ('Hot Love', 'Get It On', 'Telegram Sam' and 'Metal Guru'). After the break-up of The Beatles, he helped to forge a new, glamorous era of glitz, glitter and gender-fluid fashion, and at one point was arguably the biggest pop star in Britain. At the time of his death, Bolan was enjoying a resurgence in popularity, fronting his own successful TV show, *Marc*, which included a duet with his close friend, David Bowie, who had emerged from the folk scene at the same time as Bolan.

Fans quickly turned the tree into an unofficial shrine to Bolan. Memorial plaques and a bronze bust of Bolan also feature at the site. The tree itself has been tended by the T. Rex Action Group since 1999, when it was in danger of falling down.

The sycamore was long believed to be the direct cause of Bolan's death, but it was revealed in 2012 by an eyewitness that the car first struck the post of a chain-link fence, and it was this that killed him. Vicky Aram, a nightclub singer who had been invited home by the couple, was driving the car directly behind them with Bolan's brother Richard, and was the first on the scene. She revealed in a biography that the car only came to rest against the tree, and in fact prevented it from sliding down the hillside, sparing Jones from further injury (Jones suffered a broken arm and broken jaw in the accident).

Of course, it wouldn't be a rock star death without a conspiracy theory, and several coincidences have fuelled the suggestion that Bolan had a typically poetic premonition of his death:

- The René Magritte painting *The Sixteenth of September* depicts a tree at dawn. Jones' car came to rest against the tree at 5am on 16 September 1977.

- In the song 'Celebrate Summer', released a month before his death, Bolan sings, 'Summer is heaven in '77'.

- The number two T. Rex single 'Solid Gold Easy Action', released in 1972, contains the line, 'Easy as picking foxes from a tree'. The number plate on Jones' Mini was FOX 661L.

- In the same song, Bolan sings, 'Woman from the East with her headlights shining eased my pain and stopped my crying.' As they were driving from central to south-west London, Jones and Bolan were technically travelling from the east, being followed by Vicky Aram. Although it is believed he was killed instantly, Aram laid Bolan's body on her mother's rug, which she had in the boot of her car.

- Bolan once stated he would never live to the age of thirty; he died two weeks before his thirtieth birthday.

- One of his first ever demo tracks (recorded under the name Toby Tyler) was entitled 'The Road I'm On (Gloria)', with the refrain, 'The road I'm on won't carry me home.' At the time, Bolan hadn't yet met Gloria Jones.

I don't know if it helps to suggest there's something mystical about a man dying in a car crash. But if you want to pay your respects to an undeniable talent, you can track the tree down (on the B306 opposite Gipsy Lane) and leave your own floral tribute.

The First Bramley Tree

Apple, Southwell, Nottinghamshire, England

Unless we're talking Adam and Eve, we don't often consider the idea that a fruit might have started out with one original example. They just seem to be there, hanging off trees for us to pick and eat as we please. It's just the way nature intended.

But we do know everything there is to know about the Bramley apple (good for cooking, not so nice to eat off the tree). We know which exact tree it came from and even who planted the pip. How? Because the tree is still there, in a garden in Southwell, a short distance north of Nottingham. Lord Byron used to stay there, but for once, though, Byron isn't the most famous person in this story. That honour belongs to Mary Ann Brailsford who, as a young girl in 1809, planted some apple pips in her garden. (The date is sometimes contested but, given that they celebrated the bicentenary in 2009, we're sticking to it.)

A local butcher, Matthew Bramley, bought the house in 1846 and when, a decade later, local nurseryman Henry Merryweather noticed the potential of the tree and asked to take cuttings, Bramley agreed on the proviso that any fruit sold should be called Bramley apples (he also insisted, a little unfairly, that the original seedling planted by Brailsford be called 'Bramley's Seedling'). We even know when the first ever Bramley apple was sold – Merryweather records 'Mr Geo. Cooper of Upton Hall' bought three on 31 October 1862. Today, 95 per cent of all culinary apple orchards in England are Bramley trees.

Despite being felled by a storm in 1900, the original 'Bramley's Seedling' took root where it touched the ground. More recently, the tree has contracted an incurable fungal infection, and in 2018 Nottingham Trent University bought the rose garden containing the tree so its life can be extended and grafts can be taken for replanting.

7

The Cliveden Redwood

Redwood, Cliveden, Berkshire, England

In 1897, William Waldorf Astor, the 1st Lord Astor, imported a section of Californian redwood from Vance's Wood, Humboldt County, and had it set up in the woods overlooking Cliveden House, the English estate he bought in 1893 for $1.25 million. At just over 5m (16ft 6in) across, it is the largest section of a *Sequoiadendron giganteum* (giant redwood) in Britain.

Quite why Astor had the stump shipped all the way to the banks of the River Thames is unclear. Several theories were proposed, the most outlandish of which was that he had drunkenly entered into a $50,000 wager that trees in America were so large he would be able to find one large enough to serve as a dining table for fifty guests.

Sadly, as a report in *The San Francisco Call* from 15 January 1899 made clear, this turned out to be a malicious rumour, believed to have been spread by the captain of the British ship employed in its transportation. Astor himself was so upset at the reports that he wrote the following letter to *The Times*, also in 1899 (the same year he became a British citizen):

> Editor of the Times – Sir:
>
> Will you allow me to publish in your columns a contradiction of the reports that have been circulated about a section of California redwood recently brought to Cliveden? The section referred to has been placed on the ground as an object of interest, but it has never been intended to use it as a dining table, nor has any bet been made as to the number of persons who could be seated around it. The report repeating these details, and purporting to give an account

of a banquet, is a deliberate and mischievous fabrication. I have given instructions to my solicitor, Sir George Lewis, to commence proceedings against the newspaper, which has published the false statements in question.

Yours faithfully,
William Waldorf Astor
Cliveden, October 25.

The true reason followed Astor to his grave. Maybe he just wanted a piece of his native country to remind him of home. Although the likeliest explanation is also the simplest: he did it because he could.

Postscript
Scientists have managed to cultivate cuttings from the Fieldbrook Redwood Stump – from which Lord Astor's section was taken – which is 10.7m (35ft) in diameter. The plant nursery at the Eden Project, near St Austell, now houses ten of its clones growing as saplings as part of a plan to propagate and replant some of the oldest trees in America and Britain.

8

The Colchester Castle Sycamore

Sycamore, Colchester, Essex, England

There are many trees associated with famous historical buildings, but only one is actually growing on top of the building in question. Initially planted to commemorate the defeat of Napoleon following the Battle of Waterloo in 1815, the Colchester Castle Sycamore sits on the south-east tower of the Norman keep – itself built on top of the foundations of a Roman temple. Colchester was once the capital of Roman Britain (and the first city established by the Romans in the country) – its name denoting a special town whose residents were granted rights comparable to Roman citizens themselves.

Legend has it that the tree was planted in 1815 by the daughter of the mayor. However, some impressive sleuth work undertaken by local historians has uncovered some considerable holes in this story. The first mention of the tree doesn't occur until 1892 in a book produced by Wilson Marriage, who was mayor from 1891–2. Colchester had two mayors in 1815: John King and his successor Edward Clay. At the time, the town was beset by internal political strife, with both mayors facing moves to have them legally disqualified from office. It's possible, of course, that either mayor may have sought to curry favour from the locals by planting the tree – hoping to attract some reflective glory from Wellington's famous victory. And Colchester Castle might seem the obvious location for such a symbolic statement. However, unlike today, in 1815 the castle was in private ownership, so symbolic statements by the mayor would be harder to negotiate.

At the beginning of the nineteenth century the castle also housed the county jail – and one theory is that the gaoler, John Smith, who lived within the castle walls, may have planted the tree. He did have a daughter, Mary (aged forty-three in 1815), who was born in the castle, and may well

have been confused with the mayor's daughter over time. If it was John Smith who planted the tree, he would have had to have done it within five months of the event: the battle was fought and won on 18 June and Smith died from a broken skull sustained in an accident in November that year. Mary seems the likeliest candidate: Smith's son inherited the role of gaoler, but within twenty years – by 1835 – the gaol had closed. Mary, however, continued to live in the castle, running the library that was still housed within the grounds.

The other problem with the story is that, as I look at a contemporary photograph, the tree was too small to be a 200-year-old sycamore. You might notice the use of the word 'was' in my previous sentence. Why would I be using the past tense if the tree is still standing proudly atop Colchester Castle?

Ah. Here's the thing: in 1985, the tree was removed while repairs to the castle wall were undertaken. It transpired that, starved of nutrients, the roots of the wizened tree had spread through the walls, causing considerable displacement. Despite having its roots cut, the tree was kept alive. It was translocated to the nursery of Castle Park and replanted in its original position in May 1987. (In a neat nod to – possibly apocryphal – history, the sycamore was replanted by Nicola, the six-year-old daughter of then mayor Sir Bob Russell.)

This protracted process allowed the opportunity to analyze the tree, which certainly appeared to be the right age within the statistical parameters – its diminished size perhaps explained by its lack of nourishment while its roots grew down through the walls.

There is plenty of contemporary 'proof' both for and against the tree's authenticity – unfortunately, as it predates photography, this largely involves drawings and copperplate prints. Two illustrations from 1817 show sycamores on the site, despite there clearly being no tree in equivalent depictions from 1812 – suggesting that a more mature tree was transplanted to the castle walls rather than a sapling or it being grown from seed.

There is also evidence of subterfuge: in the 1930s, the curator of the museum within Colchester Castle chopped back the sycamore's branches, apparently hoping to kill it. But despite the attacks sustained over the last two centuries, the sycamore still thrives. Whether it stands as a symbol of anti-republicanism or a victim of local politics, it's further proof that nature outlives the petty motivations of mankind.

9

Covin Trysting Tree

Sweet Chestnut, Bemersyde,
Roxburghshire, Scotland

Dated between 500 and 800 years old (the older estimate tallying with local accounts that it was planted by Petrus de Haga in the twelfth century), the Covin Trysting Tree still stands proudly in front of Bemersyde House, on the site of ancient woodlands on the banks of the River Tweed in Scotland.

The tree – and the house – links J.M.W Turner, one of Britain's greatest artists, Sir Walter Scott, one of its greatest poets, and Earl Haig, one of its most controversial soldiers. Turner was commissioned by Edinburgh publisher Robert Cadell to produce a series of twenty-four landscape paintings to illustrate Scott's twelve-volume *Poetical Works*, published in 1833–4 (an original of which was later owned by Haig). Turner's preparatory sketches for the project, made during a tour of Scotland in 1831 at Scott's invitation, featured several houses and landscapes connected to the poet's life, including his childhood home, Smailholme Tower, and Abbotsford, the house he built and fought so hard to keep.

We know from Cadell's diary that Turner visited Bemersyde on 6 August 1831. Scott was distantly related to the Haigs, who could trace their ownership of Bemersyde directly to Petrus de Haga, who was listed in 1162 as proprietor of the lands and barony of 'Bemersyd'. The Haigs (an anglicized version of de Haga's Norman surname) remained proprietors into modern times. The estate was certainly of special significance to Scott: he was buried at nearby Dryburgh Abbey thanks to his family connection to the Haigs, and the funeral cortège paused on its way from his home in Abbotsford at a prominent viewing point on Bemersyde Hill (also sketched by Turner). This spot – a favourite of Scott's – commands views over a bend in the River Tweed, taking in Abbotsford, the Eildon Hills and the top of the house at Bemersyde. (It has been known as 'Scott's View' ever since.)

The etching of Bemersyde House used for the book was made from Turner's finished watercolour. The only sketch he is known to have made on his visit is housed within the Tate collection, and this shows that Turner added four prominent figures to this scene (as well as a waiting coach and horses). In the foreground stand Scott and Cadell alongside Mary Haig ('daughter of the Laird of Bemerside'). Behind them, Turner is depicted sketching a magnificent venerable chestnut – the Covin Trysting Tree.

It is obvious why the tree would capture Turner's attention: even then, it outdated the house by a good 400 years (the oldest parts of the current building date back to 1535). Today, the gardens appear slightly different to how they would have looked in Turner's day: that is because the current design was overseen and laid out by Douglas Haig, or – to give him his full name – Baron Haig of Bemersyde.

Haig was not a direct descendant of Petrus de Haga. His father was manager of Haig's whisky distillers and he didn't take ownership of Bemersyde until 1921, when it was purchased by 'a grateful nation' for him after it had been put up for sale by a cousin, Arthur Balfour Haig. However, for a man who was Commander in Chief of the British Expeditionary Force from 1915 until the end of the First World War, Douglas Haig's distant relatives would hardly have been what you might call champions of the British Army. Peter Haig, 6th Lord of Bemersyde, fought at the Battle of Bannockburn in 1314, while Gilbert, 11th Lord, was among those who led the Scottish army in their defeat of the English at the Battle of Sark in 1448. By contrast, William, 13th Lord, died at the hands of the English at the Battle of Flodden in 1513.

For his own part, Lord Haig is a hugely divisive figure in modern British history. First the positives: he was hailed by American General John J. Pershing as 'the man who won the war', particularly with his Hundred Days Offensive. His state funeral at Westminster Abbey on 3 February 1928 heralded a national day of mourning, with his body then escorted to his birthplace of Edinburgh, where it lay in state for three days at St Giles' Cathedral. His obituary in *The Times* for that year read: 'The greatest soldier that the Empire possessed has passed away suddenly, while still in the fullness of his powers. Lord Haig not only shouldered the heaviest military burden that any Briton has ever borne, but, when the War was over, and with the same foresight that distinguished him in his campaigns, he took up a task which probably no other could have accomplished, and devoted all his time and energy to the service of his old comrades in the field.'

However, in the years that followed, his reputation became sullied, with accusations that his headstrong, dogmatic approach and emphasis on attritional warfare had directly resulted in hundreds of thousands of unnecessary deaths. By the 1960s, thanks in no small part to unfavourable depictions in Joan Littlewood's play *Oh! What a Lovely War!* (1963) and Alan Clark's book *The Donkeys* (1961), he had come to be known by the disparaging moniker of 'Butcher Haig'.

Latterly, Haig's character has been re-evaluated: his apparent intransigence countered by his clear openness to new tactics and technologies towards the end of the war – not least the innovative 'combined arms' strategy proposed by his multinational Field Commanders (Herbert Plumer of the British Army, Canadian Arthur Currie and Australian John Monash). As for his callous indifference for human losses, Haig's supporters point out that he subsequently founded the Haig Fund and Royal British Legion (both in 1921) to support veterans of the First World War – a cause to which he dedicated the remainder of his life (and which brought him into direct conflict with both the government of David Lloyd George and King George V). The fact that Bemersyde was bought for him through public subscription speaks of the affection in which he was held by many. But, despite the twenty-nine major honours he was awarded – from Knight of the Order of the Thistle to the Croix de Guerre – the creation of his title of Earl Haig in 1919 (he became a Baron in 1921) came following considerable political friction with – and perceived snubs from – Lloyd George.

His legacy to Bemersyde, however – and its famous trysting tree – is less controversial. Despite the fact that the original trunk has now died, measures taken by Earl Haig in the 1950s have ensured that several new healthy stems now rise from its base. As for the famous picture by Turner – amongst the symbolic items placed in the foreground is a manuscript entitled 'Thomas the Rhymer'. This refers to the medieval poet who wrote the lines, 'Tyde what may whate'er betide, Haig shall be Haig of Bemersyde'. Today, the grounds of the house (and its famous tree) can be visited all year round – and it still remains in the ownership of the Haig family.

'The hill, the stream, the tower, the tree,

Are they still such as once they were?

Or is the dreary change in me?'

From 'The Dreary Change',
Sir Walter Scott

10

The Covenanter's Oak

Oak, Dalzell Estate, Motherwell, Scotland

Found on the Dalzell Estate in Motherwell – part of the original royal hunting park planted by King David I of Scotland (who reigned from 1124–54), the Covenanter's Oak is thought to be over 800 years old.

Its name, however, derives from the Scottish Covenanters of the seventeenth century, a Presbyterian movement seeking religious and civil freedom, which refused to acknowledge that the monarch was the spiritual head of the Church. Their opposition to Charles I's religious reforms helped to set in motion a chain of events that lead to the English Civil War and the king's eventual execution. But we'll save that story for other trees...

On 28 February 1638, the National Covenant was signed in the Greyfriars Kirkyard, Edinburgh, declaring opposition to the king's attempts to force English liturgical practice and church governance upon the Presbyterian Church of Scotland. It's worth noting that the wording still urged loyalty to the king, but in effect it marked a period where the Covenanters became the *de facto* government of Scotland.

However, with the Restoration of 1660 that saw Charles II – and the Stuart lineage – return to the Scottish throne, Covenanters suddenly found themselves targeted as traitors by the new king. In 1662, he renounced the covenants, declaring the oaths unlawful. The Covenanters were forced to hold their services in secret, away from their churches.

One minister, Reverend John Lawrie, having been removed from his pulpit, took to holding clandestine 'conventicles' under the great oak tree in King David's ancient royal park (with the knowledge and approval of the Hamilton family, who had been bequeathed the estate in 1645).

The services attracted congregations of hundreds, despite their attendance alone being a treasonable act. By 1680, a more radicalized group of underground Covenanters – named the Cameronians after the Reverend Richard Cameron – had openly renounced their allegiance to the king, prompting a violent escalation in their persecution. This period became known as the Killing Time: anyone caught preaching at conventicles was put to death and suspects were tortured with gruesome and barbaric methods, including being hanged by their thumbs.

The final conventicle was held under the oak in 1688, the year of the Glorious (or Bloodless) Revolution, when Mary II and William of Orange took the throne. Despite losing one of its limbs during a storm in 2008, the oak still stands today. Its appearance – cracked, split, decaying in places, its four remaining limbs propped up and braced, but still unbowed – might be a metaphor for those who met under its canopy some 300 years ago.

11

The Crowhurst Yew

Yew, Crowhurst, Surrey, England

This church yew throws up a number of mysteries. Such as: when was the door built into this ancient hollow tree? Where did the cannonball come from that was found embedded in its side? And is it really 4,000 years old – as locals claim – or a mere 1,000 years?

The Crowhurst Yew in Surrey clearly outdates the twelfth century St George's Church, which it stands next to. It's likely that the church was built in recognition of the holy site already afforded to the field in which the yew stood. The belief that Crowhurst housed a Royalist stronghold during the English Civil War might explain the existence of the cannonball, which probably became embedded in the tree as Parliamentarians fired on the farmyard opposite. This most likely would have been during a skirmish in the autumn of 1643, when Sir William Waller's Roundhead (Parliamentary) forces used Farnham, some 65km (40 miles) to the west, as a major staging post for clearing Royalists from the surrounding area.

The first record of the tree appears in 1662, when John Evelyn recorded its size as being 'ten yards in compass' in his report *Silva: Or A Discourse of Forest-Trees and the Propagation of Timber in His Majesty's Dominions*, which he delivered to the Royal Society (and published two years later, in 1664). Later, the writer and antiquarian John Aubrey measured its girth at 10 yards (9m) and an interior space of around 6ft (1.8m) in *Natural History and Antiquities of the County of Surrey* (1719). A door was added sometime after 1820 (when

'Of all the trees in England,

Oak, Elder, Elm and Thorn,

The Yew alone burns lamps of peace

For them that lie forlorn.'

From 'Trees', Walter de la Mare

villagers discovered a cannonball while excavating earth around the base of the hollow trunk) and in 1850 in *A Topographical History of Surrey* (Vol. IV), Edward Brayley described how the tree's hollow interior had been 'fitted up with a table in the centre and benches around'. Brayley also noted that, following a storm, 'The roof, however, as it may be termed, has fallen in.'

The tree is not to be confused with another yew found in Crowhurst, East Sussex (between Bexhill and Battle). Legend recalls that, shortly before the Battle of Hastings, King Harold had a local official hanged there for failing to disclose the whereabouts of a hidden supply of treasure.

12

Dering Woods

Various species, Smarden, Kent, England

The present-day allure of Britain's forests is, of course, countered by a sinister side derived from centuries of bad press. There is a fear and loathing long associated with our woodlands – historically, they were places to avoid, not to escape to. Until relatively recently it was the urban environment that offered a desirable refuge from the rustic rather than vice versa.

There are countless myths and fables to reinforce this notion. In short, scary things have always happened in woods. Haunted houses are nothing compared to the dense, dark, impenetrable masses of nature. Epping Forest in Essex, for instance, is said to be populated by the ghosts of the many murder victims buried there, as well as the perpetrators of those crimes – including the notorious highwayman Dick Turpin, whose outlaw life began there with his so-called 'Essex gang'.

But Epping Forest's claim to be the most haunted in England is challenged by the Dering Woods in Kent, also known as Screaming Woods. Situated between the villages of Smarden and Pluckley, the 160-ha (400-acre) Dering Woods encompasses one of the oldest Neolithic sites in the world, and stories of mysterious screams, whispers and footsteps being heard there at night are almost as old as the trees themselves. Among the ghosts said to haunt the woods are those of a suicidal army colonel from the eighteenth century as well as a highwayman from the same era who, it is said, was beheaded while pinned to an oak tree by angry villagers.

The spooky stories continue well into the twentieth century: on 1 November 1948, the bodies of twenty people (including eleven children) were supposedly discovered in the woods, with locals reporting seeing strange lights in the forest the previous night (All Hallows' Eve, now commonly known as Halloween). It is claimed that the autopsies were unable to determine a cause of death, and even that police involvement was suspected. However, firm proof of this story is hard to establish. On online search reveals a purported front page of *The Smarden Post* newspaper from the time, reporting the story. But it is clearly a fake. The image used to illustrate the article shows a Second World War-era Russian officer in the middle of the photograph. Investigations have proven it to have been taken in Lithuania in 1945 following the discovery of a mass grave of Jews murdered by the Nazis.

Similarly, doubt is cast over the tale of private investigator Robert Collins, who apparently died in a car accident in 1965, a year after conducting interviews about an unknown religious cult in Smarden. And the tall tales continue: of four students who disappeared into thin air in 1998; and of at least fifteen different ghosts said to have been sighted in or near Pluckley.

Not everyone is amused by such stories: locals accuse ghost hunters of desecrating the ancient forest, while the Woodland Trust has spent tens of thousands of pounds on security and clean-up operations, particularly around Halloween. Sadly, however, the area does have a long history as a popular site for suicides. And while some of the more fanciful stories have been debunked, it's still an undeniably eerie place. If you go down to the woods tonight, you're sure of a big surprise...

13

The Eynsford Arbour Day Trees

Various species, Eynsford, Kent, England

The tradition of 'Arbour Day', where communities are encouraged to plant trees in their neighbourhood, is thought to date back to 1594, when the mayor of the Spanish village of Mondoñendo introduced the ritual. Limes and horse chestnut trees still line the main square there, revealing the origins of the tradition. Reintroduced in Villanueva de la Sierra in 1805, the custom took hold in the USA in 1872, when, instigated by one J. Sterling Morton, an estimated one million trees were planted in Nebraska.

In 1897, noting the growing international movement, the town of Eynsford in Kent adopted the trend with a unique twist. A local iron merchant, Elliott Downs Till, suggested planting trees to commemorate Queen Victoria's Diamond Jubilee – but with a difference. His idea was for trees to be arranged acrostically – that is to say, with the first letter of each planted species used to spell out a sentence or phrase.

Originally planted along the main street from the railway to the town centre, the initial letters of the common name of each species of tree planted spelled out a line from the Robert Browning poem 'Rabbi Ben Ezra':

'THE BEST IS YET TO BE: THE LAST OF LIFE FOR WHICH THE FIRST WAS MADE.'

There were fifty-two trees used in the initial planting and the ones known are (with a question mark denoting those that have not been identified):

Turkish hazel
Horse chestnut
Elm (wych)

Beech
Elm
Sycamore
Thorn (American)

Ilex (holly)
Sloe

Yew
Elder
Turkish hazel

T? (possibly another
   Turkish hazel)
Oak

Beech (purple)
Eucalyptus

Turkish hazel
Hickory
Eucalyptus

Lime (common)
Acacia (false)
Sycamore
Tilia × europaea
   (common lime)

O?
Fagus (beech)

Laburnum
Ilex
Fagus
E?

Fagus
O?
Robinia

W?
Holly
Ilex
Cypress
Hickory

Tilia tomentosa
   'Petiolaris' (weeping
   silver lime)
H?
E?

F?
Ilex
R?
S?
Tilia × europaea

Walnut
Acacia (false)
S?

M?
Acacia (false)
D?
Elm (wych)

There have been many plantings since, such as those commemorating battles in the Boer War, including Kimberley, Ladysmith and Mafeking. In 1902, four years after Queen Victoria's death, thirty trees (consisting of twenty-two species) were planted, with the initial letter of each tree species spelling out a line from Alfred, Lord Tennyson's 'To the Queen' (1851):

'SHE WROUGHT HER PEOPLE LASTING GOOD'.

Sycamore	Horse chestnut	Liquidambar
Hickory	Elm	Ash
Elm	*Robinia*	Service tree
		Thorn
Walnut	Poplar	*Ilex* (holly)
Robinia	Elm	Norway maple
Oak	Oak	Guelder rose
Ulmus (elm)	Poplar	
Gingko	Laburnum	Gum
Hornbeam	Elm	Oak
Thorn		Oak
		Damson

Around the war memorial, four trees also read:
   'LOVE'

Lime	Olive	Veronica	Elm

Despite the support of the newly founded Royal Society for the Protection of Birds (formed in 1889), who suggested establishing a Bird and Arbour Day, the idea never caught the imagination of the British public as it did in the United States.

As for the fifty-two trees from the Victoria Jubilee memorial, twenty-two remain:

'... Horse chestnut, Elm; Beech, ... Thorn; *Ilex*, ...; ... ...; ... Oak; Beech, ...; ... ...; Lime, ... Sycamore, *Tilia*; **(O)** *Fagus*; ... ... *Fagus*, ...; ... ... Robinia; ... Holly, ... Hickory; *Tilia*, ...; ... *Ilex*, ... *Tilia*; Walnut, Acacia, ...; ... ... Elm.'

Of the Tennyson acrostic, however, only the initial sycamore has survived – the other twenty-nine trees having being cut down to provide playing fields for the local primary school. Which brings to mind another line by Tennyson:

'All that was left of them...
When can their glory fade?'

14

The Fortingall Yew

Yew, Fortingall Parish Church,
Perthshire, Scotland

Many claim that the ancient yew that stands in the grounds of Fortingall church, Perthshire, is the oldest tree in Europe (even at the most conservative estimate of its age, which varies wildly from 2,000–9,000 years old). Local legend has it that Pontius Pilate was born under the shade of its branches when his father served as Roman ambassador to the Caledonians.

It's more likely that Pilate was a Samnite, born in the village of Bisenti in Central Italy, although there are references in Roman literature of him spending time in Gaul and/or Germany after his time as Prefect of Judea. There is certainly evidence that some of his descendants found their way to Britain.

One version of the story has it that Pilate's father was on a diplomatic mission to a Pictish king in ancient Perthshire when news of his son's birth reached him at Fortingall; another insists that his pregnant wife had been travelling with him, explaining how Pilate came to be born here. However, the origins of the tale may be a little more recent: a hoax dreamt up by Sir Donald Currie, who bought the Glenlyon Estate in Fortingall in 1885 and was a patron to Alfred, Lord Tennyson and Rudyard Kipling. Either way, it's a good story.

Whether or not the Fortingall Yew ever sheltered Pilate (and it's certainly old enough), it holds a special place in Scottish lore, and not only for its sheer grandeur. It is said to lie in the geographic centre of Scotland, at a point where three major ley lines intersect – and was seen by Druids as a sacred tree of life or knowledge.

Today, the old yew isn't quite as imposing: the natural ageing process, as well as the attentions of souvenir hunters over the centuries, has meant its trunk has split into separate stems. However, in 1779, its circumference was recorded by Thomas Pennant as '56 and a half feet' (*The Churchyard Yew and Immortality*, Vaughan Cornish, 1946).

'Old Yew, which graspeth at the stones

That name the under-lying dead,

Thy fibres net the dreamless head,

Thy roots are wrapt about the bones.'

From, 'In Memoriam A.H.H.', Alfred, Lord Tennyson

15

The Glastonbury Holy Thorn

Hawthorn, Glastonbury, Somerset, England

The legend of Joseph of Arimathea links the man supposedly responsible for the burial of Jesus with the Holy Grail, Arthurian myth and the small Somerset town of Glastonbury. The tale, as first recorded by French poet Robert de Boron in his poem written *c.*1200, 'Joseph d'Arimathe', holds that the Grail (a drinking vessel that featured in the Last Supper), was used by Joseph to capture the last drops of blood from Christ's body while he was still on the cross. The story continues with Joseph travelling with the Grail to 'the valleys of Avaron in the west', which many have inferred to be Avalon or Glastonbury. Here, the Grail was guarded until the coming of King Arthur.

Before leaving Glastonbury across the flooded Somerset Levels by boat, Joseph is said to have struck the ground on Wearyall Hill with his staff, at which spot a Holy Thorn immediately flowered. Thus the mystery of the Glastonbury Hawthorn is apparently explained. One of very few places where it occurs, this particular hawthorn is a naturally occurring genetic mutation, which differs from common hawthorn in that it flowers twice a year – once in the spring and once on or around Christmas Day. This second flowering over Christmas is seen as a miracle, marking the arrival of Christianity in Britain. Traditionally, the hawthorn is a pagan symbol of fertility that has associations with the May Day and Green Man celebrations in Britain. Thus the idea of a new blossom coinciding with the festivities around Christ's birth holds profound significance. Glastonbury is the site of one of the most important abbeys in the country, founded as early as the seventh century AD.

Crowds would gather every Christmas Day to witness the holy flowering, but, when Britain adopted the Gregorian calendar in 1752 to bring it in line with the rest of Europe, Christmas fell eleven days earlier, meaning that, much to the consternation of those who congregated on Christmas Day, no flowers appeared. However, when they returned on 5 January, the tree was seen to flower once more, right on cue.

But the Holy Thorn's legend hasn't always been appreciated. In the 1640s, during the English Civil War, the original tree was cut down and burned by Cromwell's troops, who condemned it as a 'relic of superstition'. Even then, it was recorded that the man who swung his axe at the tree was blinded in one eye by its thorns. A new tree was planted on Wearyall Hill in 1951 on the original site and in 1965 the Queen erected a wooden cross at Glastonbury with the inscription:

'The cross, the symbol of our faith, the gift of Queen Elizabeth II, marks a Christian sanctuary so ancient that only legend can record its origin.'

The replacement tree had its branches cut off by vandals in 2010 – and its subsequent replacement sapling was further vandalized in 2012. In 2019, the landowner removed the remains entirely.

The story of the Glastonbury Thorn doesn't end there, however. Since hawthorn is commonly propagated from cuttings or graftings, there are arguably several surviving examples that can be traced back to the original tree. It is believed that the monks at Glastonbury Abbey regularly took cuttings, while in *The History and Antiquities of Glastonbury* by Charles Eyston (1722), it is stated, 'There is a person about Glastonbury who has a nursery of them, who, Mr Paschal tells us he is informed, sells them for a crown a piece, or as he can get'. A hawthorn in the grounds of Glastonbury Church as well as the Church of St John were both said to be graftings of Holy Thorn cuttings onto the roots of blackthorn stock – and further cuttings have been exported as far afield as New Zealand, Australia, the USA and Canada.

The Great Fraser Yew

Yew, Beinn a' Bhacaidh,
Inverness-shire, Scotland

In woodland overlooking Loch Ness from its south side, stands the old hollow trunk of the Great Fraser Yew, surrounded by its offspring, which form a grove almost 30m (100ft) in diameter. It gives the impression of a great clansman itself, calling a meeting of its devoted followers – which is apt. The Clan Fraser of Lovat settled in Inverness-shire in the thirteenth century, probably from France, and took a yew sprig as their symbol.

Before battle, the Frasers would meet around this great tree (which would have been magnificent and ancient even then), taking a piece of yew to wear in their bonnets. It is known that they gathered here before marching to Culloden. Sir Simon Fraser ('The Patriot') fought alongside John Comyn III of Badenoch, William Wallace and Robert the Bruce. He was captured and executed by Edward I on 1306, but his cousin, Sir Alexander Fraser, fought at Bannockburn and later became Chamberlain of Scotland.

The Frasers were involved in several famous sieges of Inverness: the first to support Mary, Queen of Scots in 1562 (see pp.125–129); and again in 1649 and 1650 while opposing the parliamentarians. By 1715 however, they were supporting the British government against the Jacobites. It didn't last. Affronted at being stripped of his captaincy by George II, Simon Fraser, 11th Lord Lovat switched his allegiance and the Frasers of Lovat fought in the front lines of the Jacobites in the Battle of Culloden in 1746.

Today, the Fraser name is common across North America (many settled there after fighting the French at Quebec) and in Australia and New Zealand – both of which have had a Fraser prime minister (Malcolm Fraser (1975–83) and Peter Fraser (1940–49) respectively). Wherever Frasers now settle, they still are joined by the symbol of the sprig, taken from the grand old tree on the hillside above Loch Ness.

The Green Gage Tree

The Gage family have a long and storied past, contributing greatly – for better and worse – to England's history over the centuries. For instance, Sir John Gage KG (1479–1556) served under four monarchs in total (no small feat in itself given the lifespan of the average lord at court in the sixteenth century).

Sir Thomas Gage (c.1719–87) created and trained the first light infantry in the British Army and went on to fight both alongside and against George Washington. His ally against the French during the Seven Years' War in Canada (1756–63) soon became his greatest adversary, when Gage was made Commander in Chief of the British Forces in North America in 1763, charged with supervising the thirteen American colonies. It didn't end well. His heavy-handed response to the Boston Tea Party in Massachusetts prompted the Battles of Lexington and Concord – in effect kick-starting the American War of Independence. He was sent home ignominiously in 1775, generally held responsible for the rise in tensions that led to Britain losing its prize colony. (His response to the demand for greater self-governance was to write that 'democracy is too prevalent in America'.)

Another Gage, Sir William, 7th Baronet (1695–1744), was Member of Parliament for Seaford from 1722 until his death and, more importantly, was one of the earliest patrons of cricket. His Firle team, still based within the grounds of the family home, Firle Place in Sussex, developed a fierce but friendly rivalry with Charles Lennox, 2nd Duke of Richmond's neighbouring Sussex team. A game between the two in 1725 is one of the earliest recorded cricket matches, thanks to Gage's letter to Lennox on 16 July of that year:

'I am in great affliction with being shamefully beaten Yesterday, the first match I played this year. However I will muster up all my courage against Tuesday's engagement. I will trouble your Grace with nothing more than that I wish you Success in everything except ye Cricket Match...'

But we are not interested in the origins of cricket or the light infantry here, but rather those of a fruit that bears the family name. And it was another William Gage who first introduced the 'Green Gage' to these shores in 1725 – the very same year as that pioneering cricket match.

The man responsible was Sir William Gage, 2nd Baronet of Hengrave, from a different branch of the family. He was a cousin and contemporary of his namesake (living from 1650–1727), and lived in Hengrave Hall, near Bury St Edmunds in Suffolk. Two years before his death, he imported the fruit – a cultivar of the plum – from France, where it is known as the *Gross Reine Claude* ('large Queen Claude', after Queen Claude, Duchess of Brittany). Its origins are rather more exotic, and can be traced back to the *goje sabz*, a distinctive, green-coloured wild plum from Iran. (Iran is still the world's seventh largest producer of plums, while other species of plum have been traced to China and Eastern Europe.)

Rather neatly, greengages were soon after introduced into the American Colonies – with greengage trees grown on the plantation of George Washington himself. So the Americans can thank the Gage family not only for their freedom... but their plums.

18

The Hagley Wych Elm

Scotch Elm, Hagley Wood,
Worcestershire, England

On 18 April 1943, four boys were hunting for birds' nests in the woods of the Hagley Hall estate in Worcestershire when they chanced upon a gruesome find. Searching within the hollow trunk of a large wych elm (or Scotch elm), one of the boys, Bob Farmer, discovered a human skull.

Obviously spooked – and worried that they would be in trouble for trespassing – Farmer and his three friends (Fred Payne, Robert Hart and Thomas Willetts) replaced the skull and returned to their homes. Willetts told his parents what they had found, and the police were alerted.

Upon examining what had now become a crime scene, the police discovered that the elm's trunk concealed almost an entire skeleton (the remains of a hand were also found close to the site). The body was in a relatively early stage of decomposition: tufts of hair, a section of flesh and fragments of clothing were still apparent, prompting the forensic examiner to propose it might have been left in the tree as recently as eighteen months before its discovery. Furthermore, the body would have had to have been squeezed into its final resting place fairly soon after the time of death before rigor mortis set in.

Little more is known about the victim, other than that the body was of a female, aged between thirty-five and forty, and that a section of taffeta material – thought to be part of her underdress – was found in her mouth. A cheap, imitation wedding ring was also found on her finger.

Beyond that, there is mystery and obfuscation.

Rumours abounded: was she a prostitute murdered by a client – hence the fake wedding ring? Did the severed hand, the taffeta in her mouth and the choice of 'wych' elm point to an occult ritual? Or – as some attest – was she a Nazi spy?

To add to the confusion, in 1944, six months after the discovery of the body, a graffiti message appeared on the nearby eighteenth-century

Hagley Obelisk on Wychbury Hill. The message asked 'Who put Bella in the wych elm?'

The same message has appeared frequently but sporadically on the monument ever since, but no likely candidate for 'Bella' has been found, despite searches of dental records. Efforts to identify her have been compounded by various circumstances: firstly an overwhelmed police force during the war, when there was a surfeit of missing persons cases; and latterly the fact that at some point 'Bella's' skull disappeared from the police station at Tally Ho, Pershore Road, Birmingham (where it had been kept since it was first moved from the forensics lab of Professor James Webster, who carried out the post-mortem).

Perhaps the most intriguing – and fanciful – theory was that 'Bella' had either stumbled upon – or been covertly involved in – a Nazi spy ring during the war. This is backed up by the declassified files of Josef Jakobs, an agent for Abwehr (Germany's military intelligence during the Second World War) – and the last man to be executed in the Tower of London (on 15 August 1941). Upon breaking an ankle while parachuting into Cambridgeshire, he was arrested by the Home Guard, who discovered a photograph of a woman about his person. Jakobs claimed this was his lover, a German actress and singer named Clara Bauerle, who had spent time before the war in Birmingham, had perfected a local accent, and was being trained as a spy. The Home Guard was reprimanded at the time for leaking details of the case – could they have conflated the name 'Bauerle' into 'Bella'?

Regardless, there is no evidence that Bauerle ever returned to the UK, and reports detail that she actually died in Berlin in December 1942. So if not her, then who? Another candidate is Clarabella Dronkers, a Dutch agent again embroiled in a German spy ring that included Birmingham's dance halls.

Both German spy explanations earned greater credence following an anonymous letter to the *Express & Star* in 1943, claiming the murdered woman had been passing on information about munitions factories in the Birmingham area.

With each account seemingly more fanciful than the last, perhaps the original explanation – that Bella was a prostitute murdered by a punter – is the most plausible. In 1953, a local woman named Una Mossop made a statement to police that her late ex-husband, Jack Mossop, had confessed to the manslaughter of a woman he had met in the Lyttelton Arms in Hagley along with a Dutchman named 'Van Ralt'. After she had passed out from drink, Mossop claimed the two men had placed her comatose body inside a hollow trunk in the woods in the hope that she would wake up in the morning and learn 'the error of her ways'. Furthermore, he was said to have died in an asylum, haunted by visions of a woman in a tree, before the body had even been discovered.

However, Una Mossop – the only corroborating witness for her own story – chose not to share any of this information until ten years after her ex-husband's death. With yet more doubt clouding the truth, almost eighty years later the question still remains unanswered: who put Bella in the wych elm?

'Full in the midst a spreading elm display'd

His aged arms, and cast a mighty shade,

Each trembling leaf with some
light vision teems,

And heaves impregnated with
airy dreams.'

Virgil, *The Aeneid*, circa 30-19BC

19

The Hardy Tree

Ash, St Pancras, London, England

Nestled in the churchyard of St Pancras Old Church in London is a curiosity: among the tombs of the great and the good of eighteenth- and nineteenth-century London (Sir John Soane, Mary Wollstonecraft – mother of Mary Shelley – and Baroness Burdett-Coutts), is a solitary ash tree, around which are laid hundreds of gravestones, set perpendicular to the tree and extending outwards in rings, many of them consumed by roots and foliage.

The name 'The Hardy Tree' relates not to the relative resilience of the ash tree and its roots, but to the assumed author of its design. This peculiar monument dates back to the 1860s, when plans for the new Midland Railway at King's Cross ran across an area of the church's cemetery. The work of designing the new line was assigned to the architect Arthur Blomfield (1829–99), a proponent of the Gothic Revival style and famed for his designs for Covent Garden. As for the unpleasant job of overseeing the disinterment and reburial of many of the remains in the churchyard, Blomfield delegated that to his young assistant, a promising young architecture graduate from Dorset... by the name of Thomas Hardy.

Before his fame as one of the pre-eminent writers of Victorian Britain, Hardy (1840–1928) had first trained under a local architect in Dorchester. In 1862, at the age of twenty-one, he moved to the capital, enrolling in evening classes at King's College London where he studied modern languages. As Mark Ford notes in his biographical work *Thomas Hardy: Half a Londoner*, Hardy made something of a name for himself, winning several prizes including a silver medal from the Royal Institute of British Architects (RIBA) for his essay 'On the Application of Coloured Bricks and Terra Cotta to Modern Architecture', and a first prize from the Architectural Association for his designs for a country mansion.

'If woods be suffered to be felled,

as daily they are,

there will be none left.'

King James I

It's not certain how directly Hardy was responsible for laying the tombstones around the tree that bears his name, or if it was the artistic eye of one of his underlings who came up with the striking pattern. What we can be sure of is that, in 1865, Hardy was placed in charge of the excavation and reburial work by Blomfield, who he was assisting at the time. As such, he would undoubtedly have been aware of the work, and most likely instructed it to be carried out. What we can also be sure of is that he noted the gruesomeness of the task in his diaries and, by 1867, had returned to Dorset, apparently after being laid low by some form of nervous exhaustion.

Having first resumed work as an architect with his original employer in Dorchester, Hardy finally turned to a life in literature. His first novel, *The Poor Man and the Lady*, was finished in 1867, the same year he returned to the West Country, but failed to find a publisher. Eventually, however, with the publication of *Far From the Madding Crowd* in 1874, Hardy was able to give up architecture, going on to become one of England's most celebrated authors. But this was not before leaving us with an unusual and macabre monument in the capital, with a backstory he would surely have appreciated...

20

The Hiroshima Tree

Cherry, Tavistock Square Gardens,
London, England

At 8.15am local time on Monday 6 August 1945, a 4,400kg (9,700lb) uranium-enriched gun-type atom bomb, nicknamed Little Boy, was dropped from the USAAF Boeing B-29 bomber *Enola Gay* above Hiroshima, Japan. It exploded at a height of 600m (almost 2,000ft) over the city, with an energy of 15 kilotons, effectively flattening an area of 3.5km (2 miles) in diameter. The shockwave of the explosion was accompanied by a fireball with a surface temperature of 6,000°C (11,000°F). This fireball and the subsequent radioactive fallout was estimated, in 1946, to have killed around 140,000 people (a number that was revised to 202,118 in a 1998 study taking into account deaths from longer-term radiation-induced cancers that could be linked to the explosion).

Little Boy was one of only two nuclear bombs ever used in warfare. The second, Fat Man, was detonated just three days later over Nagasaki (a back-up target after the original, Kokura, was obscured by smoke from a previous firebombing raid). This time a plutonium implosion-type device was used – but the effects were just as apocalyptic. The 21-kiloton explosion resulted in a total of 60,000–80,000 fatalities.

It's impossible to imagine this degree of devastation. Leonard Cheshire, a group captain in the RAF who acted as an official British observer of the Nagasaki bombing, later wrote:

> 'By the time I saw it, the flash had turned into a vast fireball which slowly became dense smoke, 2,000 feet above the ground, half a mile in diameter and rocketing upwards at the rate of something like 20,000 feet a minute. I was overcome, not by its size, nor by its speed of ascent but by what appeared to me its perfect and faultless symmetry...

'Loveliest of trees, the cherry now

Is hung with bloom along the bough,

And stands about the woodland ride

Wearing white for Eastertide.'

From *A Shropshire Lad*, A.E. Housman

"Against me," it seemed to declare, "you cannot fight." My whole being felt overwhelmed, first by a tidal wave of relief and hope – it's all over! – then by a revolt against using such a weapon.'

Cheshire was not alone in feeling revulsion for the horrors that humankind can inflict upon one another. The justification for unleashing such weapons on civilians relied on cold mathematical calculations about the number of casualties on both sides that an invasion of mainland Japan might incur. While the bombings undeniably hastened the Japanese surrender, and thus the end of the Second World War, the atrocities suffered by the citizens of Hiroshima and Nagasaki raises the question of whether the end justified the means.

In the intervening years, there have been countless memorials around the world to those who died in those two atomic bomb blasts. In Tavistock Square Gardens in London's Bloomsbury, a flowering cherry tree was planted in memory of the Hiroshima victims on 6 August 1967 by Councillor Mrs Millie Miller, Mayor of Camden. It was an apt location for such a tribute: in 1920 the Tavistock Clinic was founded on the square to treat psychiatric patients, including those suffering from shell shock following the First World War. The site is now occupied by the headquarters of the British Medical Association (BMA).

The year after the Hiroshima tree was planted, in May 1968, Prime Minister Harold Wilson unveiled a statue by sculptor Fredda Brilliant of Mahatma Gandhi (1869–1948), leader of India's non-violent independence movement and a prominent champion of civil rights.

The square has come to be known as the Peace Garden due to the number of trees, statues and monuments it holds dedicated to the pursuit of peaceful initiatives. The large Conscientious Objectors' Commemorative Stone ('To all those who have established and are maintaining the right to refuse to kill. Their foresight and courage gave us hope') was unveiled near the square's North Gate entrance by the Peace Pledge Union on 15 May 1994 to mark International Conscientious Objectors' Day.

A sculpture of the writer Virginia Woolf (1912–41), a prominent member of the Bloomsbury Group of literary figures, thinkers and artists, stands in the south-west corner of the square (placed there in

June 2004 by the Virginia Woolf Society of Great Britain and based on a 1931 original by the artist Stephen Tomlin). A *Ginkgo biloba* dedicated to Virginia's husband, author Leonard Woolf (1880–1969), was planted there in December of the same year. As well as being notable figures in the pacifist movement, the couple were residents of 52 Tavistock Square, living there between 1924 and 1939; the building was subsequently destroyed following a bombing raid during the London Blitz in October 1940. Opposite Virginia's sculpture is a memorial designed by the architect Sir Edwin Lutyens dedicated to Dame Louisa Aldrich-Blake, Britain's first female surgeon (1865–1925), who is renowned not only for her early work treating cervical cancer, but also for performing surgery on the front line during the First World War.

———

At 9.47am on Thursday 7 July 2005, eighteen-year-old Hasib Hussain set off an explosive device on the number 30 double-decker bus he was riding on as it passed Tavistock Square. In total, thirteen people were killed in the explosion, one of a series of co-ordinated suicide bombings across the city by converts to the Islamist group al-Qaeda that killed a total of fifty-two people. It is a grim irony that a square dedicated to pacifism and the remembrance of past atrocities should itself be visited by such murderous destruction.

In September 2018, a memorial stone was laid in the square, bearing the names of the victims of the 7/7 bus bombings, with London Mayor Sadiq Khan amongst the attendees. It replaced a temporary plaque that had been fastened to the railings opposite the BMA building (some of whose staff were among the first to arrive on the scene to deliver assistance to the injured).

It feels only appropriate to end with a quote from the man whose likeness sits beatifically in the middle of Tavistock Square:

'The weak can never forgive. Forgiveness is an attribute of the strong.' Mahatma Gandhi

21

The Kilravock Castle Kissing Beech

Beech, Kilravock, Inverness-shire, Scotland

There are countless examples of trysting trees across the country (see pp.51–54). Strictly speaking, a trysting tree is one that is popular for any type of liaison. Thus the Great Fraser Yew (p.74) could be said to be one example, since the Fraser clan have historically met there. Similarly, the Covenanter's Oak (pp.56–57) was a regular gathering place for religious sermons. But, in common understanding, trysting trees are used for one purpose: romance.

The name of the Kilravock Castle Kissing Beech brooks no ambiguity: it's here for the kissing. It is also the largest example of a layering beech to be found in Scotland: a magnificent example, its trunk almost 5m (16.4ft) in girth, its draping limbs sprouting forth in all directions, snaking to the ground where they have taken root. In fact it is better to describe it as a number of trees all with the same parent.

Beeches are particularly suitable as trysting trees – not just for the number of low-lying, ground-hugging boughs to sit upon, but also because their smooth, grey bark makes a perfect surface for lovers to carve their initials, hearts and other symbols of devotion into. The Kilravock beech bears the names and memories of many who embraced here. But its name comes from one particular incident, when the son of the local laird – possibly a future owner of Kilravock himself – was caught in a clinch at the tree with one of the housemaids from the castle.

There is little physical evidence to give firm names or dates to this tale – certainly no initials have been attributed to the pair – but it is a story repeated over the centuries by the many lovers who have subsequently met here, illicitly or not.

Kilravock Castle was built in 1460 and has been continuously owned and lived in by members of the Rose family, who were originally French nobles who arrived in England following the Norman invasion, settling at Kilravock in 1293. A continuous line of direct descendants of the first Laird Rose have been resident ever since, entertaining some of the most important figures in Scottish history, including Mary, Queen of Scots, Robert Burns and Bonnie Prince Charlie.

As for the identity of the mystery laird-to-be with the wandering hands: for lack of any better theories, there is a higher than usual probability that his first name was Hugh, as it has been a common forename among the Rose family's firstborn sons, stretching back for centuries. Hugh Rose then: caught with his maid under the beech tree, K.I.S.S.I.N.G.

22

The Kingley Vale Great Yew

Yew, Chichester, West Sussex, England

There are two things every schoolchild remembers about the Battle of Agincourt that took place in 1415 between the English army led by Henry V and the French. First, that the peculiarly English habit of 'flicking the Vs' – sticking up the index and middle finger in a V-shape as a gesture of rudeness or defiance – originated with the king's archers. Rumours spread that the French army had threatened to cut off the first two fingers of any archers they captured so that they would never be able to fire arrows again. Having helped to secure an overwhelming victory, despite being vastly outnumbered, it is said that the archers then taunted their captured foes by sticking their still-intact fingers up at them. (Another, rather more prosaic, theory is that the 'V' simply signified the number five as they celebrated King Henry.) Second, that the archers brandished longbows hewn from one of England's ancient yew forests – Kingley Vale in West Sussex.

Sadly, both stories appear to be apocryphal. It's true that, in his rousing pre-battle speech, Henry V warned his longbowmen that the French had threatened to cut off their fingers. However, Jean de Wavrin, the French chronicler who was an eyewitness at the battle, records that Henry specifically mentioned their first *three* fingers would be cut off (since English longbows are operated with three fingers). Furthermore, there is no contemporary evidence of the gesture being used in the fifteenth century, other than the odd medieval tapestry depicting someone brandishing two fingers, but with no clues as to their intention. In fact, the first unequivocal evidence of the gesture being used as an insult isn't found until 1901. This occurs in early documentary footage of the Parkgate Iron and Steel Company, Rotherham (available to view for free on the British Film Institute website). Here, as workers line up to enter the works, a young man gestures aggressively to the camera with two fingers.

As for Kingley Vale, it is indeed one of the most ancient and grandest yew forests in Europe. Some of the yews growing there are believed to be among the oldest living organisms to be found in Great Britain. But English yew is notoriously unsuitable for making longbows. Far from being the supple wood required for bowmaking, it is actually brittle – as well as being riddled with knots. Instead, the English archers of Agincourt preferred the straight-grained varieties of yew found in Spain and Portugal – and royal decrees exist showing they were imported from those countries by the king, either direct or via Venice, often in boats also carrying shipments of wine.

But that is not to diminish the majesty of the Great Yew of Kingley Vale on the South Downs. One of Britain's oldest trees, it stands within the site of an ancient Druidic ceremonial meeting ground, and is close to the four Bronze Age burial mounds known as the Devil's Humps. One legend suggests the Great Yew was planted in AD 894 as a memorial for a battle against Viking raiders – an incident mentioned in G.N. Garmonsway's *The Anglo-Saxon Chronicle* (1953):

> 'When the host that had besieged Exeter sailed back on
> its way home, it harried inland in Sussex near Chichester,
> but the garrison put them to flight and slew many hundred
> of them, capturing some of their ships.'

Whatever legend or myth we choose to believe or debunk, the Great Yew remains a magnificent specimen – its gnarled, twisted boughs reaching out like outstretched fingers, producing an enveloping canopy. The yew is often seen as a symbol of immortality, and whether it was there to witness druid rites or Viking battles, Kingley Vale remains a place that inspires mystery and awe.

23

The Knole Oak and the Strawberry Fields Tree

Oaks, Knole Park, Kent, England

The convoluted story of the seven oaks that lent their name to a town in Kent (see pp.146–147) is further muddied by two other famous trees associated with Knole Park in Sevenoaks.

Now owned by the National Trust, the history of Knole estate – with an area of nearly 400ha (1,000 acres) – dates back over 600 years. The land was bought by Thomas Langley, Bishop of Durham, in 1419 and the existing house can trace its origins back to 1456, when Thomas Bourchier, Archbishop of Canterbury, acquired the estate and began renovations there. Knole House remained the property of successive archbishops until debts forced Thomas Cranmer to sell it to Henry VIII in 1537. The estate ended up in the hands of Thomas Sackville, a cousin of Elizabeth I, in 1603 (the year of her death).

Countless generations later, Knole was the childhood home of Vita Sackville-West, the writer and pioneering garden designer. Sackville-West was the barely disguised inspiration behind her lover Virginia Woolf's masterpiece *Orlando* (a perennial contender whenever someone draws up a list of 'the greatest British novels').

In the early pages of the book, Orlando is described walking 'very quickly uphill through ferns and hawthorn bushes, startling deer and wild birds, to a place crowned by a single oak tree.' This oak becomes a pivotal, recurring motif in the book, and inspires *The Oak Tree*, the epic poem that the titular character begins as a boy. Indeed, when Sir Nicholas Greene finally comes across the manuscript of Orlando's epic, after an interval of some 300 years, his verdict seems to summarize Woolf's own expectations of perfection in writing:

> 'There was no trace in it, he was thankful to say, of the modern spirit. It was composed with a regard to truth, to nature, to the dictates of the human heart, which was rare indeed, in these days of unscrupulous eccentricity. It must, of course, be published instantly.'

Since Orlando's home is undoubtedly modelled on Knole, this oak could well be based on a genuine 'special' tree from Sackville-West's childhood. There are certainly several venerable contenders, old enough to assume the role – the most likely being the tallest sessile oak in Britain, which still stands in the park at a height of 41m (134ft). (Although it is perhaps not, as Woolf describes, 'so high indeed that nineteen English counties could be seen beneath; and on clear days thirty or perhaps forty' – unhelpfully masking its identity behind the air of unreality that shrouds the whole book.)

There is, however, another far more identifiable oak tree that stands in the grounds of Knole Park. This one is far easier to track down, since it has been filmed: it appears in the promotional video for 'Strawberry Fields Forever' by The Beatles, filmed in 1967. In the video, the tree, a dead oak located behind Knole's Bird House, appears as an extension of a bizarre stringed instrument that, at one point, Paul McCartney jumps up into backwards (taking a somewhat literal approach to the line, 'no one I think is in my tree...').

The Beatles also filmed parts of their video for 'Penny Lane' at Knole at the same time – and it's remarkable that the only two singles in which they specifically name-check locations from their Liverpool upbringing should be accompanied by visuals filmed 400km (250 miles) away in Kent. Knole, incidentally, plays another important role in the story of The Beatles. While making the two films, John Lennon wandered into an antiques shop in Sevenoaks on 31 January 1967 and bought a nineteenth-century poster advertising 'Pablo Fanque's Circus Royal' at Town-Meadows, Rochdale. The text of the poster formed the basis of the lyrics to the song 'Being for the Benefit of Mr. Kite!', which the band began recording just a few weeks later, on 17 February 1967.

24

The Leith Hall Dule

Sycamore, Aberdeenshire, Scotland

Among the 106ha (260 acres) making up the grounds of Leith Hall, Aberdeenshire, stands an impressive sycamore, known to locals as the Dule Tree. The word *dule* derives from the Gaelic for 'grief'; which will give you some indication as to the tree's former sinister purpose.

Dule (or 'dool') trees were used for execution by hanging, and also often as a gibbet for displaying the corpse of the executed man as an example to others. Sycamores were commonly used as dules due to the strength of their branches. The Leith Hall Dule is believed to date from around 1650, when the Leith-Hay family built their stately pile on the site of the medieval Peill Castle. Although the sycamore was probably planted at the time the new hall was built, there was likely to have been a gallows tree at the castle previously. Baronies in Scotland retained the ancient tradition of the 'pit and gallows' right up until the Heritable Jurisdictions Act of 1746, which sought to bring the vagaries of local justice under official government legislature.

Several ghosts are said to haunt the site, including that of Laird John Leith III, who died after being shot in the forehead in a drunken duel on Christmas Day, 1763. Numerous guests have reported seeing his apparition with a bandaged head while staying at Leith Hall – including several patients who were treated there during the First World War, when the building was used as a temporary hospital.

Due to their gruesome role, dule trees are often found in prominent positions – at a crossroads or on the grounds of the local landowner's estate. Other notable examples include the Blairquhan Castle Sycamore, on the Blairquhan Estate in South Ayrshire, thought to have been planted by the Kennedy family around 1500, during the reign of James V of Scotland.

Ash trees – such as the Logierait Ash in Perthshire – were also used as dules, due to the durability of the wood and its ability to absorb pressure

without splintering. Contemporaneous accounts record how the ash at Logierait was 'the dool tree of the district, on which caitiffs and robbers were formerly executed, and their bodies left hanging till they dropped and lay around unburied'.

Another can be found beside the remains of a ruin dating from the thirteenth century in South Lanarkshire. The castle itself was the inspiration for Sir Walter Scott's 1831 novel *Castle Dangerous*, the last published in his lifetime. Scott, remembering Douglas from his youth, revisited the site while working on his book in order to refresh his memory of the castle, as well as researching local traditions and customs. (In another of Scott's books, *Guy Mannering* (1815), he refers to a 'Justice-Tree' at the fictional Ellangowan Castle.)

As one might expect from the name, Tom Nan Croiche (the 'Hill of the Gallows'), which stands above the village of Dalmally in Argyll and Bute, was a prominent spot for execution. The tree that now stands at the site is not the original dule – which at some point was replaced by a more traditional gallows: the hole where its post was set can still be seen here.

There are several notorious examples of gallows trees south of the border too, of course.

The Heddon Oak, between Stogumber and Crowcombe in Somerset, was perhaps the most infamous example. It was here that six of the victims of Judge Jeffreys' 'Bloody Assizes' were hanged for their part in the Monmouth Rebellion of 1685. After hanging, they were tarred and then quartered in the field behind the oak. The tree, on a crossroads in the Quantock Hills, was once said to be haunted, with locals reporting hearing the sound of heavy breathing and its boughs creaking from the weight of bodies. Sadly, it creaks no more, having been cut down by overzealous tree surgeons, claiming (erroneously) that it was rotten. Locals, however, still refuse to walk past the crossroads at night.

The Llangernyw Yew

Yew, Conwy, Wales

One of the unique characteristics of yew trees is their ability to survive in 'fragmented' form, even if the original core has died off. Ancient yews in such a state are variously described as split, shattered, decayed, ruined or wrecked, yet they live on: a phenomenon explaining how yews are associated both with immortality and, conversely, death.

The ancient yew in the churchyard of St Digain's in the village of Llangernyw, Conwy, North Wales is a prime example. It has fragmented into several large offshoots, its core lost (indeed the space left between the trunk fragments was so large that it once housed the oil tank providing power for the church). This has resulted in some wildly divergent estimates as to its age, varying from 1,500–5,000 years old (if only it was as easy as counting rings). It's worth noting that, at the most generous estimate, the Llangernyw Yew would probably qualify as the oldest tree in Britain – arguably the world. But that alone isn't enough to afford its place in history here.

According to local folklore, the Llangernyw Yew is home to the Angelystor: which translates in Welsh as the 'Recording Angel' or 'Evangelist'. The Angelystor is said to appear twice a year – on the evening of Halloween and on the 31 July – on which dates it loudly and solemnly announces the names of those parishioners who will die in the following year. Of course many doubt the existence of the spirit, not to mention its sinister predictive powers. But they might be well advised to keep their opinions to themselves: legend states that, one Halloween, a tailor by the name of Shôn ap Robert openly mocked the idea of Angelystor while drinking in the local pub. However, when – egged on by his fellow drinkers – he visited the churchyard afterwards to put forward his bold claim, he was met by a booming voice calling out his own name. According to contemporary reports, he replied by shouting, 'Hold, hold! I am not ready yet!'

Within the year, it is recorded, he was dead.

The Major Oak, Nottingham

Oak, Sherwood Forest,
Nottinghamshire, England

> Robin Hood:
> 'Where lies that cask of wine whereof we plundered the
> Norman Prelate?'
> Little John:
> 'In that Oak where twelve can stand inside nor touch each
> other.'
>
> *The Foresters*, Alfred, Lord Tennyson (1891)

It may not be one of his most revered works, but there is little doubt that Tennyson was thinking of the Major Oak when he wrote the above lines for his 'play with incidental music' (alternatively titled *Robin Hood and Maid Marion*).

Often referred to as Britain's favourite (and most photographed) tree – and indeed voted England's inaugural Tree of the Year in 2014 – the Major Oak is estimated to be between 800 and 1,000 years old.

When we talk of 'venerable' trees, this is surely the image that one conjures up in one's mind: at once grand but decrepit; formidable and gnarled, yet hollow inside and artificially propped up out of deference to its vast, sagging limbs. With a trunk circumference of 10–11m (33–36ft), a canopy spread of 28m (92ft) and weighing roughly 23 tonnes, the Major Oak is also the biggest oak tree in Britain. It may be the result of multiple saplings that have fused together over the centuries, or else it may have been pollarded at some point, encouraging the trunk to grow thick and wide. It has certainly taken its time to reach such an impressive size; its position within a clearing has allowed it to spread its

canopy widely without competition from neighbouring trees, while the relatively poor soil where it stands in Sherwood Forest (near the village of Edwinstowe in north Nottinghamshire) means it would have grown slowly and steadily, adding to its strength.

Having just read the words 'Sherwood Forest', you will probably have guessed what's coming next. Folklore inextricably links the Major Oak with the legend of Robin Hood. The famous outlaw of Sherwood Forest is said to have lived here with his Merry Men during the reign of King John between 1199 and 1216. Despite its extraordinary size, the Major Oak is not so named because of its girth. Rather, it is named after a local historian, one Major Hayman Rooke who, in 1790, wrote a book detailing the oak trees of the area named *Remarkable Oaks* (or, to give it its full title, *Descriptions and Sketches of Some Remarkable Oaks in the Park at Welbeck in the County of Nottingham, a Seat of His Grace the Duke of Portland, to Which are Added, Observations on the Age and Durability of That Tree, with Remarks on the Annual Growth of the Acorn*).

If the upper estimate of the tree's age is correct, the Major Oak would have been a mature 200-year-old tree during the reign of King John: if the lower end is accurate, it is unlikely it would have supported the Merry Men in its boughs. The forest itself certainly held other trees of the right vintage: in his book, Rooke notes that during felling, oaks were found bearing the branding marks of King John buried deep within their rings.

Before Rooke's time, the tree was known locally as the Cockpen Oak: the area was famous for a breed of cocks, and the tree's hollow trunk provided a perfect site for them to roost. Whether Robin Hood, Little John, Friar Tuck *et al* also found shelter there is a moot point. It's notable that another of the Major Oak's contemporaries was a huge, cave-like oak known as 'Robin Hood's Larder' – where the Merry Men were supposed to have stored their supplies (including poached game). However, this, like many of the great old oaks of Sherwood, only survives in old pictures and local anecdotes, having succumbed to storms, fires or natural decay.

To protect it from a similar fate, the Major Oak has been fenced off since the 1970s, protecting its roots from the footsteps of visitors and preventing any budding Will Scarlets and Alan-a-Dales from clambering in its boughs. It is partially supported by an elaborate system of scaffold-like supports, first installed at the same time as the fencing, and updated in the 2000s. Chains were attached to its crown in 1908.

Of course, all this conjecture ignores one simple fact: the legend of Robin Hood is exactly that, a legend. The first mention of him appears in the fifteenth-century ballad 'Robin Hood and the Monk', and it seems likely that the tales of his life are an amalgamation of various stories, real and imagined. He has been variously described in early English poetry as Robyn Hude (*c.*1420), Robyn Hod (*c.*1475) and Robyn Hode (*c.*1592). The surnames Whod or Whood also appear in the seventeenth century. The names of Robin Hood and Maid Marion occur in several Shakespeare plays, too (*The Two Gentlemen of Verona*, *As You Like It* and *Henry IV Part I* and *II*). As for existing historical figures – there are several possible candidates, including: Robert Hod of York, an outlaw mentioned in legal papers in 1225; Robert Godberd, an outlaw from the 1260s; Robert Hood of Wakefield, a rebel opposing Edward II in 1322; Robin of Loxley (from the sixteenth century) plus the catch-all name 'Robehod' for any outlaw in the thirteenth century. (Indeed, some suggest that the term 'Robin Hood' itself became shorthand for anyone on the wrong side of the law, who might typically seek their hiding place in England's dense forests – hence the confusion of identity between several individual stories.)

Whether he hailed from Nottinghamshire or Yorkshire – or whether he lived at all, Robin Hood will forever be associated with the Major Oak. Either way, as Tennyson will attest, it makes for a great story. And anyone holed up in Sherwood Forest or just visiting over the past millennium couldn't help but look in awe at what is arguably Britain's greatest tree...

Newton's Tree

Apple, Woolsthorpe, Lincolnshire, England

One of the best-known urban (or semi-rural) myths in science tells us how Sir Isaac Newton formulated his law of universal gravitation after watching an apple fall from the tree under which he was resting. And yet it's more than mere legend: according to contemporary reports, it's a story that Newton himself told.

Voltaire (in his *Essay on Epic Poetry*, 1727) first records how, while Newton was walking in his garden in either 1665 or 1666, he 'had the first thought of his System of Gravitation, upon seeing an apple fall from a tree.' Furthermore, the tree in question still exists, in the garden of Woolsthorpe Manor, Newton's birthplace and family home, near Grantham in Lincolnshire.

In his *Collections for the History of the Town and Soak of Grantham* in 1806, Edmund Turnor identifies the specific tree in question as a rare Flower of Kent variety, noting 'The apple tree is now remaining and is showed to strangers.' Strangers are still shown the tree today, courtesy of the National Trust, who now own the property and present it in as close a representation as possible to the converted yeoman's farmstead that Newton lived in from his birth on Christmas Day, 1642.

Having just completed his BA at Trinity College, Cambridge, Newton was embarking on an MA in 1665–6 when the university was forced to close due to the Great Plague epidemic. Newton was compelled to return to his childhood home to complete his studies in optics and the refraction of light (he is credited with building the first practical reflecting telescope as well as developing groundbreaking theories on colour and the visible spectrum, collated in his book *Opticks* of 1704). While there (as he related to his friend, the antiquarian and Stonehenge expert William Stukeley, for his 1752 biography of Newton), 'the notion of gravitation... was occasion'd by the fall of an apple, as he sat in contemplative mood.'

This simple, everyday occurrence – something Newton must have witnessed hundreds if not thousands of times at Woolthorpe, prompted him to question (as he recalled to Stukeley) "'Why should that apple always descend perpendicularly to the ground?"... "Why should it not go sideways, or upwards? But constantly to the earth's centre? Assuredly, the reason is, that the earth draws it. There must be a drawing power in matter. & the sum of the drawing power in the matter of the earth must be in the earth's centre [sic]...'"

Of course, being a Cambridge scholar of physics, his enquiring mind didn't end there. Some two decades later, he asked, 'How far would that power extend?... Why not as far as the moon?' This was the guiding principle behind his 'universal' theory of gravity, as published in his *Philosophiæ Naturalis Principia Mathematica* (1687), which extrapolated from his apple theory and the astrological theories of Kepler to explain how gravitational pull holds the solar system in orbit.

All of which means – if its identification is correct – that the Woolsthorpe tree holds a distinguished place in the history of modern thought. It is certainly old enough – aged at around 400 years old, it would have been a mature tree of forty or fifty years old at least at the time of Newton's 'Eureka!' moment. Despite being blown down in 1820 (not long after Turner had positively identified it), the tree took root where it fell and continues to flourish and bear fruit.

A descendant of Newton's apple tree grows beneath his lodgings in Trinity College, Cambridge, outside the main gate. In recent years, cuttings have been donated to Loughborough University with the aim of creating 'clones' of Newton's tree, and in 2015, pips from the tree were carried into space by British astronaut Tim Peake, while on the Principia mission on board the International Space Station. Having spent six

'The tree which moves some to tears of joy is in the eyes of others only a green thing that stands in the way. Some see nature all ridicule and deformity... and some scarce see nature at all. But to the eyes of the man of imagination, nature is imagination itself.'

From a letter to Dr Trusler,
William Blake, 18 August 1799

months floating in microgravity, they were taken to Wakehurst Place in Sussex (the gardens and seed bank managed by the Royal Botanic Gardens, Kew), where under constant monitoring they have since been nurtured into 'space saplings'. Neatly combining the fruits of Newton's theories on gravity with the fruit that inspired them, it's proof that the apple never falls far from the tree, even when it's been into space and back...

28

The Oak at the Gate of the Dead

Oak, Wrexham, Clwyd, Wales

With its vast trunk cleft in two, this ancient oak more than lives up to its intimidating name. It is thought to be over 1,000 years old, dating back to the time of King Egbert (or Ecgberht) of Wessex, who reigned from AD 802–39.

The tree didn't split until 2010, finally defeated by centuries of weathering and the cycle of rainwater seeping inside and expanding as it freezes. Until then, the trunk had an impressive 10-m (33-ft) girth. It stood as an imposing gateway at the head of the wooded Ceiriog Valley, some 300m (almost 1,000ft) below Chirk Castle, on Offa's Dyke, the eighth-century earthwork that follows the border between England and Wales.

It was in this valley that Welsh forces commanded by Owain Gwynedd ambushed Henry II's invading English army at the Battle of Crogen in 1165. The dead were buried close to where they fell, in the ditch of Offa's Dyke – a site that came to be named Adwy'r Beddau (literally 'the pass of the graves'). The 'gateway' is actually created by the space formed between another impressive tree, some 7–8m (23–26ft) in girth. Legend tells of a local duel fought beneath its boughs, though the names of those involved – or indeed details of where the loser was buried – have been lost in the mists of time.

Today, the oaks mark the gateway to nothing more sinister than a pleasant country walk, but its name is a reminder of the medieval warriors buried in the valley.

'Then, dearest Maiden! Move along these shades

In gentleness of heart; with gentle hand

Touch – for there is a spirit in the woods.'

From 'Nutting', William Wordsworth

29

The Ormiston Yew

Yew, Ormiston Hall, West Lothian, Scotland

This is the tree where, according to legend, the seeds of Scottish Reformation were sewn. It was from the natural 'cathedral' of boughs radiating from this ancient yew that John Knox and his mentor, George Wishart, preached in the sixteenth century.

Knox (*c.*1514–72) was born near Haddington, a small town in East Lothian. He was ordained in 1536 and became a tutor to the sons of local lairds. However, he became increasingly influenced by the Protestant teachings of Martin Luther, much of it smuggled from England, which fuelled a growing contempt for the corruption of the Catholic clergy. With a passionate zeal, Knox began to preach of 'purifying' the Scottish church. His fiery oration drew little distinction between reformation and violent revolution.

As such, the Reformists were regarded as a danger by the establishment, which attempted to stifle the spread of heresy and insurrection. Wishart himself was arrested in 1546 and, in a show trial orchestrated by Cardinal Beaton, Archbishop of St Andrews, was sentenced to death. He was hanged from a gibbet and then burned at the stake at St Andrews. (Beaton was subsequently murdered by a group of enraged Protestant nobles – an act that Knox explicitly approved of.)

A year later, Knox himself was arrested, but was eventually released. Taking the hint, he lived in exile, preaching first in England and then travelling through Europe. He met the French reformer John Calvin in Geneva. It is clear the two were greatly impressed by each other's zeal. His fiery opinions often brought confrontation even among his so-called allies – in Frankfurt, his headstrong approach prompted a group of Protestants to storm out of the church, refusing to be in the same building as him.

He was equally provocative with the pen, with titles including his *Faithful Admonition* and *The First Blast of the Trumpet Against the Monstrous*

Regiment of Women, in which he accuses Mary Tudor of being 'a traitoress and rebel against God.'

Returning to Scotland in 1559, the itinerant firebrand became a lightning rod for Protestant disaffection. Becoming a minister in Edinburgh, his congregations would become pumped up by his rhetoric, engaging in riotous acts that targeted symbols of the Catholic church.

Eventually, in 1560, under the Protestant Elizabeth I, the Treaty of Berwick was signed. Parliament approved a Reformed Confession of Faith (drafted by Knox), which effectively declared the abolition of the 'old faith' and denied papal jurisdiction in Scotland.

In his 'Lament for the Earl of Hopetoun' (1824), James Miller imagines the Yew of Ormiston speaking to its audience:

> 'Here Wishart shew'd prophetic powers,
> Before that vial of wrath was given,
> When in St Andrew's dungeon towers
> His vengeance-blood uprose to heaven!'

There is no mention of Wishart's equally famous disciple. The Earl of Hopetoun's former home, Ormiston Hall, now lies in ruins. But the Great Yew, which has witnessed the beginnings of earldoms, religions and parliaments germinate beneath is boughs, still looks on.

Oswald's Tree

Ash, Oswestry, Shropshire, England

On August 5, AD 641, the pagan king Penda of Mercia amassed his own forces, together with the Welsh armies of Gwynedd, Powys and Pengwern, against Oswald, the Christian king of Northumbria. They met at the town of Maserfield in modern day Shropshire, where Oswald was killed. In the Venerable Bede's *Ecclesiastical History of the English People* (731), it is recorded that Oswald died a martyr's death and 'ended his life in prayer' for the souls of his soldiers, before he was decapitated by an axe and his body dismembered by Penda's men.

Legend has it that, as a warning to others, Oswald's limbs were hung from the branches of an ash tree by his victors, and that a holy well sprang up at the spot where his arm fell from the tree. Variations suggest that one of his arms was carried to the tree by an eagle, from where it then dropped to the ground below.

The town of Oswestry ('Oswald's tree') has stood on the site of Maserfield since at least 1190. Oswald's Well marks the spot where the spring is supposed to have originated. Oswald was made a saint for converting the people of Northumbria to Christianity, and his remains are said to have been removed to St Oswald's Priory in Gloucestershire in 909 by Lady Aethelflaed (the formidable wife of Aethelred II, the last king of Mercia), who gathered together St Oswald's various relics so that they could be interred in one place.

The Peace Tree

Oak, Stonehaven, Aberdeenshire, Scotland

Compared to many of the venerable and ancient trees mentioned in these pages, the Peace Tree at Stonehaven is a sprightly teenager, born at the start of the twentieth century. The oak was planted at Dunnottar Church on 6 July 1919 as a symbol to mark the ending of the First World War. The ceremony was led by Reverend D.G. Barron and attended by local dignitaries, including members of the Masonic Lodge of Stonehaven No. 65. The planting itself was carried out by Provost Greig, while two boy scouts stood beside him holding standards.

The tree now has a preservation order from Aberdeenshire Council and is recognized as one of the few official 'living' national war memorials to the dead of the First World War.

For a period it was left engulfed by brambles but has been better attended in recent years, and on 6 July 2019, to commemorate its 100th anniversary, a new service was held, echoing the first. Again, Brethren from the Lodge of Stonehaven No. 65 were in attendance, having kept the original spade used to plant the tree, displaying it above a door in the Masonic Lodge.

The most famous living monument to the fallen of the First World War is the National Memorial Arboretum in Staffordshire: a 60-ha (150-acre) woodland site consisting of 30,000 trees, all planted specifically for the act of remembrance, with 350 memorial plots dedicated to an individual or an organization. It is also the only site in England where a daily Act of Remembrance takes place, with the Last Post and Reveille played at 11am in the Millennium Chapel of Peace and Forgiveness.

Whether a mighty forest or a single oak, trees serve as dignified reminders of lives lost: representing a place of sanctuary, somewhere to contemplate in peaceful solitude. Quoted in the book *Tommy's Ark* by Richard van Emden, one survivor of the First World War, Lieutenant Richard Talbot Kelly of the 52nd Brigade Royal Field Artillery, recalls:

'To me, half the war is a memory of trees: fallen and tortured trees, trees untouched in summer moonlight, torn and shattered winter trees, trees green and brown, grey and white, living and dead. They gave names to our roads and trenches, strongpoints and areas. Beneath their branches I found the best and the worst of war: heard nightingales and smelt primroses, heard the scream of endless shells and breathed gas; rested in their shade, spied from their branches, cowered in their roots. They carried our telephone lines, hid our horses, guided us to and from battle and formed the memorial to many efforts of arms.'

32

The Pembroke Dock Ginkgo

Ginkgo, Pembroke, Pembrokeshire, Wales

Situated in the Royal Dockyard at Pembroke, this Japanese ginkgo tree marks a little-known link between Wales and Japan.

In 1875, Jacobs Pill shipyard in Pembroke Dock was commissioned to build one of the first modern warships for the fledgling Imperial Japanese Navy. The ship, a Kongo-class ironclad corvette named the *Hiei*, was completed in 1877. Its construction was supervised by its eventual captain, Lieutenant Togo Heihachiro, who was already in England as one of eleven Japanese officer cadets selected to further their naval studies at the Royal Naval College, Greenwich.

During his frequent visits, Togo stayed at the Master Shipwright's house and remarked on the kindness shown to him there. While attending the launch ceremony of the vessel with the Japanese Ambassador, Jushie Wooyeno Kagenori, they presented the shipyard with a Japanese ginkgo sapling as a sign of their gratitude.

Togo returned to Yokohama in Japan in command of the *Hiei* in 1877, and went on to complete a long and distinguished naval career, later becoming an Admiral and Commander in Chief of the Combined Fleet, overseeing the defeat of the Russian navy at Tsushima during the Russo-Japanese War of 1904–5 (a catalyst for the mutiny on the battleship *Potemkin* later in 1905). Today, he is revered as arguably Japan's greatest naval hero and referred to as the 'Nelson of the East'.

The bonds between Pembroke Dock and the Japanese navy were further strengthened when, on 4 October 1918, the 8,000-tonne Japanese cargo liner the *Hirano Maru* was torpedoed by a German U-boat while 51.5km (32 miles) off the coast of Pembrokeshire, with the loss of 292 lives. Some of the bodies that washed ashore were buried in

local village cemeteries, and the victims are still honoured there today, with a new memorial recently erected in the churchyard of St Mary's in Angle.

The original ginkgo presented by Togo still thrives in the garden of the Master Shipwright's house where it was planted, over 140 years ago. Saplings grown from cuttings are being nursed in the National Botanic Garden of Wales, and will eventually be shipped to Japan to be planted in places of honour in the many cities with connections to Togo.

33

The Reformers' Tree

Oak or Elm (destroyed either way),
Hyde Park, London, England

The story of the Reformers' Tree is peculiar among the entries in this book in that (a) it no longer exists and (b) no one seems to agree on its location, or even what species it was.

The initial evidence seems pretty clear-cut: a circular black-and-white mosaic set in the north-east corner of Hyde Park serves as a plaque to the legendary tree. Spiralling around the edge of the mosaic is an inscription carved in stone, which reads:

> 'This mosaic has been designed to commemorate the "Reformers Tree", a venerable tree which was burnt down during the reform league riots in 1866. The remaining stump became a noticeboard for the political demonstration and a gathering point for reform league meetings. A new oak tree was planted by the then Prime Minister James Callaghan on 7 November 1977 on the spot where "Reformers Tree" was thought to have stood.'

So the mosaic (dating from 2000, unveiled by Labour MP Tony Benn) commemorates a tree that stood somewhere else. There is also no indication of where the 'new oak tree', planted in 1977, can be found. There are no obvious candidates nearby, and no plaques to identify it. But the confusion doesn't end there.

Certainly, the Reform League, which campaigned for all adult men to be given the vote, was one of the many political groups to congregate in Hyde Park (which became a meeting point for social protest in the nineteenth century). While modern reports state that an oak tree became the focus of their protests in 1866, an eyewitness report from former police sergeant Edward Owen contradicts this.

Owen spent twenty years at Hyde Park police station (still to be found nearby, in the centre of the park) and, in 1875, recalls that the tree was 'I believe an elm like its neighbours... but not a vestige of green or anything to indicate that species simply a stark, blasted-looking old trunk, dead as a doornail, whether from lightning or old age, it had fallen into such a state, I am unable to say, but that is how it appeared in the year 1875, and was recognized as "The Old Reform Tree". The occasion of its demolition, or the cause of it, happened at a meeting or demonstration in the summer of the year mentioned above. It was not a political meeting, but a trade grievance, and I remember very largely attended. So far as the meeting was concerned it had gone off orderly and quiet, resolutions had been passed, and people were really dispersing homewards. I may add it was on a week-day, and took place in the evening, I presume to give employees every facility to attend; however, it was getting dusk, when suddenly smoke and sparks were seen issuing from the old tree, and it became apparent it had been set on fire, and that we conjectured, by mischievous boys; burn and smoke it did alarmingly, for it was nothing more than a lump of tinder... We could do nothing without water to put the fire out... We cleared the crowd back some 20 yards from the smouldering tree until the arrival of a small manual fire engine, brought by a couple of firemen, arrived and soon put an end to any sign of fire, and the crowd finally dispersed. To prevent a repetition of a similar scene, the Park authorities soon decided to have it removed altogether. Still there is the space where the old tree stood, if any of my readers care to take a walk and see as I have described...'

So, it may have been an elm and not an oak (although contemporary illustrations seem to back up Sergeant Owen's account that it was an elm) – and it may have been burned down by 'mischievous schoolboys' rather than reformists. But the shady facts surrounding both its origin and demise shouldn't deter from its lasting impact. The principles of freedom of speech, dissent and civil disobedience that the Reformers' Tree symbolized were subsequently formalized by an 1872 act of parliament allowing public speaking (as long as no indecent or obscene language was used) in the north east corner of Hyde Park – later named Speakers' Corner...

34

Rizzio's Chestnut

Chestnut, Dalkeith, Midlothian, Scotland

You will notice that most of the tales recorded in this book relate to men. The misogyny of history is too vast a topic to tackle in a book dedicated to trees – and yet it rears its head even when the tale involves two of the most powerful and charismatic women in Britain's long and winding story.

First the short story. Found near the banks of the North Esk on the grounds of the Melville Castle Hotel near Dalkeith, this stunning chestnut tree was planted for Mary, Queen of Scots by her Italian secretary David Rizzio, as a token of his love for her. Such a blatant and public gesture inevitably prompted rumour and disproval, eventually resulting in his murder. Even by the standards of the age, he met a particularly violent end. On 9 March 1566, while Mary was enjoying an evening meal at Holyrood Palace with Rizzio and other friends, a group of around eighty Scottish nobles – led by Mary's second husband, Lord Darnley – stormed into the dining room. As a shocked Mary (seven months pregnant with Darnley's child at the time) was held at pistol point, one of the nobles, Lord Ruthven, announced that Rizzio had offended her honour. In the ensuing tumult, Ruthven drew a dagger and stabbed Rizzio, who was then led away. Rizzio was stabbed a total of fifty-six times and his lifeless body thrown down the stairs.

To understand the circumstances behind this and subsequent events, we need to dig a little deeper into the gruesome story of Mary, Queen of Scots. It is, I should warn you, a convoluted one, full of twisted plotting (and plot twists). It also involves a confusing number of people called Mary or Henry.

Let's begin with the basics: Mary, Queen of Scots – or Mary Stuart – (1542–87), inherited the Scottish throne at the age of six when her father, King James V, died. As Mary I of Scotland, she is not to be confused with Mary I of England, aka Mary Tudor or 'Bloody Mary' (1516–58), who was

the eldest child of Henry VIII by his first wife, Catherine of Aragon. Mary Tudor was succeeded in England by Elizabeth I, the daughter of Henry's second wife, Anne Boleyn, and Mary Tudor's only surviving sibling. Mary Stuart herself had a strong claim to the English throne, being the great-granddaughter of Henry VII of England – and was supported in this by her father-in-law, Henry II of France.

Things get more complicated. Mary Stuart was living in France when she acceded to the Scottish throne – having been betrothed to Henry II's son, Francis. In 1559, Francis and Mary became King and Queen of France when Henry II died in a jousting accident. Francis died just a year later due to an abscess on the brain, and so the line of succession went to the ten-year-old Charles IX, Francis' younger brother, who was crowned the new King of France.

Mary returned to Scotland in 1561, and on 29 July 1565 she married for the second time, taking as her groom Henry Stuart – aka Lord Darnley. He seemed a poor match. Various reports describe him as a violent drunk, a narcissist, vain, arrogant, a conspirator and of having 'unreliable qualities' (as if the above weren't proof enough). Darnley, incidentally, also had his own claim to the English throne – as his name suggests, he was distantly related to his new bride.

It soon became apparent that the marriage was one of political convenience for Darnley as much as Mary. Despite the pregnancy, the couple become estranged. Mary, frustrated by Darnley's chauvinistic attitude – he was behaving as if the title of king made him superior to her – pointedly had a new batch of coins minted, bearing the slogan 'Marie and Henry... Queen and King of Scotland', reversing the previous (traditional) wording.

By this time, Mary had already begun her dalliance with Rizzio, who had arrived in Scotland from Turin, originally as a valet to the ambassador of the Duke of Savoy. An accomplished musician, he stayed on in court, ostensibly to sing in the choir, but was soon appointed Mary's confidential secretary and 'decipherer' in 1564.

The theory of the cuckolded husband taking vengeance against his wife's lover doesn't entirely stack up, however. Darnley's ambition involved interfering with Scotland's delicate religious balance. (He was born into Catholicism, but increasingly showed favour towards the (Protestant) Scottish Reformation – even refusing to take the nuptial mass with Mary after their wedding.) Intriguingly, he included Rizzio as a confidante in this plotting. There has even been conjecture that the two men became lovers themselves.

Then, another twist. In keeping with his character assessment above, Darnley began an audacious scheme that would see him become lawful king in his own right, with the support of parliament, and also the return of the Scottish lords exiled to England for supporting the Reformation. To absolve him of blame for acquiescing to Mary's Catholic reign, he spread rumours that Rizzio was an agent of the Pope, sent by Rome to seduce Mary and whisper sweet papacy in the Queen's bedchamber.

In the end, all the plotting came to naught for Darnley. The conspirator of unreliable qualities was himself murdered on 10 February 1567, eight months after Mary had given birth to their son, James (who went on to become King James I of England and VI of Scotland). The cause of death was inconclusive: while recuperating from a case of smallpox in a house near Holyrood, two barrels of gunpowder exploded under his bedroom. His body – dressed only in a nightshirt – was found outside, alongside that of his valet. There were no visible signs of struggle: the likelihood being he died of internal injuries from the explosion or was smothered at some point before or after the detonation.

The Earl of Bothwell – who became Mary's third husband on 15 May 1567 – was implicated in Darnley's murder, and proved to be every bit as bad as his predecessor. When Mary was returning from Linlithgow Palace in West Lothian soon after the murder, he intercepted her cortège with a force of 800 men, insisting that danger awaited her at Edinburgh and that she should accompany him to his castle at Dunbar. Some allege that he in fact took her hostage, and even that he raped her. Letters were later discovered purporting to reveal her illicit support for Darnley's killing, and public opinion began to turn against Mary and her new king. Bothwell was exiled and imprisoned while Mary sought refuge in England, gambling on the support of her cousin Elizabeth I, but instead of receiving a kinswoman's welcome, Mary was held in a succession of prisons.

Things then went from bad to desperate. Further intercepted letters implicated Mary in a failed plot to assassinate Elizabeth. A show trial followed, and among Mary's many protestations, perhaps the most compelling was her claim that she couldn't be accused of treason, since she wasn't an English subject, but rather the queen of a separate nation. Mary was found guilty and condemned to death. Without Elizabeth's knowledge, her chief advisor, William Cecil, ordered that the sentence be carried out immediately. Mary was beheaded at Fotheringhay Castle on 8 February 1587. Despite protocol demanding that monarchs should be afforded due deference, it wasn't a clean execution: it took three blows of the axe in all – the first striking the back of her head.

Perhaps things would have been different if Mary could have found a husband who loved her as much as David Rizzio. The Dalkeith Chestnut serves as a reminder of what might have been, as do the five oak trees that line the drive of the estate. Less celebrated perhaps, but equally as symbolic, they were planted by Mary in response to Rizzio's gesture of undying love, and still survive defiantly to this day.

35

The Robert Burns Trysting Tree

Hawthorn, Millmannoch,
South Ayrshire, Scotland

'At length I reached the bonnie glen,
Where early life I sported,
I pass'd the mill and trysting thorn,
Where Nancy aft I courted
Wha spied I but my ain dear maid
Down by her mother's dwelling!
And turn'd me round to hide the flood
That in my een was swelling.'
'The Soldier's Return', Robert Burns

Arguably Scotland's greatest poet, Robert Burns would often walk along the Mannoch Road that joins Mauchline and Dalrymple in East Ayrshire. On one occasion, John Thom, the local miller at the Mill of Mannoch near Coylton, recalled seeing Burns standing on the road, taking in the view across the fields. From here, his walk continued across a footbridge over the Water of Coyle, taking him through a glen and past a hawthorn tree that was a popular trysting spot for locals.

Not far beyond stood Bankhead – now long since abandoned – which Burns describes in 'The Soldier's Return' as being where Nancy's mother lived.

You won't find a more forensic description of how to find the tree than the one mentioned in Burns' ballad poem. When you get there, however, something might look awry. Sure enough, a hawthorn tree stands at the location described by Thom, surrounded by iron railings. But it is far too young to have stood on the same spot in 1793 when Burns wrote about it.

The original hawthorn died in 1916, a sad end in a time when more soldiers' returns were desperately hoped for, and its role as a trysting spot had severely diminished. The miller at the time, James Pearson Wilson, had the old tree cut down and used the wood to make small plaques and gavels, each bearing an inscription of the great poet's words. The tree found at the site today was grown by Wilson from seedlings taken from the original tree described by Burns.

A cross-section of the tree is held by the Burns Scotland Partnership, while other examples of Wilson's handiwork are housed in museums and societies around the world, from Australia to Mexico.

36

Robert The Bruce's Yew

Yew, Loch Lomondside, Scotland

In 1306, having seized the Scottish throne, Robert the Bruce and his ragtag guerrilla army of just 200 men were in retreat when they found themselves on the wrong side of the vast waters of Loch Lomond.

Robert was reeling from two defeats in quick succession – against the English at Methven Wood near Perth, and at Dalry – against forces seeking to avenge the murder of his rival to the Scottish throne, John Comyn III of Badenoch (aka Comyn the Red). Needing to cross to the western shore of the loch, Robert's men scoured the banks until one of his most loyal allies, Sir James Douglas, discovered an abandoned and damaged rowing boat.

After repairing the leaking boat as best they could, the men found it could carry only three of them at a time. Robert and Douglas were first to cross with an oarsmen, disembarking just north of Firkin Point. Here, Robert took shelter under an ancient yew growing out of a rocky outcrop.

In all, it took almost 24 hours to ferry the 200 men across the loch. As his army slowly gathered, Robert raised their spirits, entertaining them with songs, jokes and tales of valour. At times he turned to the 2,000-year-old tree and talked of its endurance, growing through the rocks as it did on a windswept hillside, drawing parallels with their own struggles.

The army took the story to heart, wearing a symbol of the yew on their uniforms, and in 1314 Robert won independence for Scotland, defeating Edward II's English army at Bannockburn. Yew trees also supplied the wood for his archers' longbows – many harvested from Loch Lomondside. (Indeed, the island of Inchlonaig is said to have been planted with yews by Robert the Bruce especially for his Royal bowmen.)

Despite 'The King's Tree' recently being found to be over 50 per cent decayed, recent remedial work, including cutting back neighbouring trees to allow more sunlight to reach it, means it remains a symbol of endurance to this day.

37

The St Lawrence Lime

Lime, the St Lawrence Ground,
Canterbury, Kent, England

There are many reasons to love cricket.

For a start, there are few other games where you can watch a single match for five days solid and still celebrate the fact there's no result. This is a sport that inspires batsmen to selflessly walk before being given out, while accepting mental disintegration as a legitimate tactic. But rather than going into all the game's many and various idiosyncrasies, let's just concentrate on one: the St Lawrence Lime.

Faced with a tree growing in the middle of a pitch, any other sport would do one of two things: move the pitch, or chop the tree down. But cricketers think differently. So, when Kent County Cricket Club's home ground was founded in Canterbury in 1847 (then called Beverley, but latterly changed to the St Lawrence Ground), it was built around a local lime tree.

Obviously, having a tree the wrong side of the boundary can pose a few practical problems, but Kent's players soon settled on the rule that a shot striking any part of the tree was worth four runs: a bonus if the ball trickled into its substantial undergrowth, but a bit annoying if the tree's 37m (120ft) high branches obstructed what would otherwise have been a six. (The only player since the Second World War to successfully score a six over the tree was Carl Hooper of Kent and the West Indies, in 1992.) Another rule deemed that batsmen could not be given out if caught from a rebound off the tree.

Sadly, in January 2005, high winds finally brought the tree down. However, the club had already been working on a successor, since the St Lawrence Lime had been found to be suffering from heartwood fungus. In March 2005, the fledgling tree was moved within the playing field, but this time outside the boundary.

The only other tree known to stand within the boundary of a first-class cricket ground is the oak at the City Oval ground in Pietermaritzburg, South Africa – the tree dates back to 1888 and its planting is thought to have been directly inspired by the St Lawrence Lime. The venue has hosted two international matches in its lifetime, both at the 2003 World Cup – and observes another tree-related tradition: notable players are invited to plant a tree near the ground to commemorate their careers, including Malcolm Marshall, Sachin Tendulkar and Sir Alastair Cook.

38

The St Maelrubha Oak

Oak, Isle Maree in Loch Maree,
Gairloch, Scotland

A 'wish tree' is any tree that, due to the belief that it holds a special religious, spiritual or superstitious significance, is presented with offerings in the hope that individual wishes or prayers will be granted. These offerings – to gods, spirits, saints or nature itself – can be in the form of anything deemed of value to the relevant culture, from food and drink, to 'treasures' of various kinds (including, historically, animal or even human sacrifices).

The St Maelrubha Oak on Isle Maree is one of the best examples of a 'coin tree', its bark glistening with coins that have been hammered into the wood. These coins serve as votive offerings and are generally associated with pleas to cure various illnesses and ailments, particularly madness (the waters of a nearby well are said to have healing properties).

St Maelrubha (AD 644–722) is believed to have lived as a hermit on Isle Maree (which is referred to as St Maelrubha's Island by locals). Also known as Máel Ruba, Maol Rubha, Máelrubai or Malruibhe (and sometimes by the Latinized name Rufus), the Irish saint is venerated as one of the apostles of the Picts in Scotland. Having lived and studied in Bangor in present-day County Down, he founded the monastic community of Applecross in the west of Ross, opposite the islands of Skye and Raasay in Scotland. He dedicated himself to an evangelical life, attempting to convert the Pictish pagans to Christianity. His missionary work saw him travel through the Highlands and Islands as well as Sutherland. In total, twenty-one ancient parishes across Scotland are associated with his name. But St Maelrubha's Island remains a key focal point: the holy well dedicated to him was frequented by those seeking its curative powers up to the eighteenth century despite the influence of the Reformation (see pp.114–115).

'Need by need by need into its hale

Sap-wood and bark: coin and pin and nail

Came streaming from it like a comet-tail.'

Seamus Heaney, 'The Wishing Tree'

Maelrubha's name is still invoked against madness and mental illness, and it was common for 'lunatics' to bathe in the waters of the well. Among the common treatments was for the sick to drink water from the well, followed by taking a dip three times a day in the loch's water, for a total of three weeks. Alternatively, they were fastened behind a boat and rowed around the entire island. It is also recorded that bulls were sacrificed in the saint's name well into the eighteenth century. Indeed many of these superstitions were still prevalent when Isle Maree was visited by Queen Victoria in 1877. It is still said that any visitor who leaves the island taking anything originating there with them (even a pebble from the shore) will be visited by madness.

While much of the superstition surrounding the island is no longer taken seriously, the tradition of hammering a coin into the St Maelrubha Oak was popularized by Queen Victoria, who mentions it in her diaries from the time of her visit, when she stayed at the Loch Maree Hotel at Talladale (resulting in the local waterfall being renamed Victoria Falls). Ironically, while visitors still continue the practice to this day, they now press the coins into the stump that is all that remains of the tree – which eventually died from copper poisoning.

39

The Sagranus Stone

Dolerite stone, St Dogmaels,
Pembrokeshire, Wales

Considering this is a book about trees, you may have noticed something out of the ordinary about this particular entry. It's a stone.

But let's not allow petty details to ruin a good story. The parish church of St Thomas the Apostle can be found next to the ruins of the earlier St Dogmaels Abbey in the village of St Dogmaels (or Llandudoch in Welsh), on the west bank of the Teifi estuary in Pembrokeshire. In the western nave of the church stands the Sagranus Stone. Thought to date back to the fifth century AD, this 2m (7ft) inscribed pillar stone is made of dolerite or diabase, which today is commonly crushed for use as aggregate for road, rail and other construction work. It is also used for headstones, memorials and as an ornamental stone – indeed bluestone, a form of dolerite, was one of the materials used in the building of Stonehenge. For some historians, the Sagranus Stone is considered just as important. That's because, much as the Rosetta Stone provided the key to translating Egyptian hieroglyphics, so the Sagranus Stone was crucial in translating the early medieval Irish language of Ogham.

The marks on the face of the imposing menhir-style slab are in both Latin and Ogham – also known as the Language of Trees, because each individual letter corresponds to a species of tree. (Hence its relevance here.) There are thought to be around 400 surviving inscriptions of Ogham on stone artefacts – mainly in Ireland (particularly in the southern province of Munster), but there are many examples in Pembrokeshire, the county that sits on the most south-westerly promontory in Wales.

The fact that the two languages are inscribed side by side on the same stone meant that it was able to act as a cipher, showing how they correspond to one another. The Latin inscription, running vertically down the stone, reads: 'SAGRANI FILI CVNOTAMI' (denoting that the stone is dedicated to 'Sagranus, son of Cunotamus' – a local chieftain). The Ogham characters translate as: 'SAGRANI MAQI CVNATAMI'.

Given the dedication, one theory suggests that the stone served as a grave marker for Sagranus. The fact that the stone isn't local, and predates not only the church (which was rebuilt several times) but also the abbey (which dates from the twelfth century and stands on a former monastic site that was raided by Vikings in AD 988), suggests that it was moved more than once before arriving at its present location. There is further evidence that it was used both as a footbridge and a gatepost in the seventeenth century, before its significance was recognized.

As for that significance: the distinctive linear marks of Ogham that run vertically down the length of the stone hint at a cross-cultural purpose, cutting across socio-political divides, possibly in an attempt to unite Celtic and Roman families of importance.

This slightly contradicts one theory: that Ogham was devised as a cryptic Irish alphabet to keep communications secret from Roman invaders (as well as 'Romanized' Britons). Another theory, proposed by the Irish scholar and archaeologist R.A. Stewart Macalister, is that the language was based on the Greek alphabet and was originally relayed both orally and via hand signals, before being transcribed as notation first on wood and later stone.

This suggestion gained prominence thanks to Robert Graves, the First World War poet turned classicist – best known for his novel *I, Claudius* – who, in his 1948 polemic *The White Goddess* proposed an all-encompassing theory of a matriarchal poetic muse that informs most early religions. Key to his hypothesis was the Celtic Tree Calendar, drawing on the Ogham alphabet and its association with tree species. Graves proposed that individual Ogham characters corresponded with ancient Greek hexameter (lines of poetic verse) revealing an all-knowing tree goddess that he names Druantia. Druantia is posited as the Queen of the Druids, her name derived from *drus* or *deru* – the Celtic for oak – and represents the 'eternal mother', the goddess of fertility in both nature and humankind.

As impressive and all-encompassing as Graves' theory is, it has largely been discredited – even Macalister himself was largely dismissive of it. His tree calendar doesn't correspond to any Celtic equivalent, and Druantia is a fictional invention by Graves.

There is no doubt that the twenty original letters of Ogham (or, to give the language its proper name, *Beith-luis-nin*) correlate to species of tree or bush. The alphabet – together with each letter's Old Irish name and the tree it relates to – is listed below:

T	*Beith*, or birch
⊤⊤	*Luis*, or rowan
⊤⊤⊤	*Fearn*, or alder
⊤⊤⊤⊤	*Sail*, or willow
⊤⊤⊤⊤⊤	*Nion*, or ash
⊥	*Uath*, or white-thorn (hawthorn)
⊥⊥	*Dair*, or oak
⊥⊥⊥	*Tinne*, or holly
⊥⊥⊥⊥	*Coll*, or hazel
⊥⊥⊥⊥⊥	*Ceirt*, or apple
/	*Muin*, or vine
//	*Gort*, or ivy
///	*nGéadal*, or fern/reed
////	*Straif*, or blackthorn
/////	*Ruis*, or elder
+	*Ailm*, or pine
++	*Onn*, or furze/gorse
+++	*Úr*, or heath/heather
++++	*Eadhadh*, or aspen/poplar
+++++	*Iodhadh*, or yew

A fine example of an ancient *Iodhadh* can be found immediately outside the south door of St Thomas the Apostle, facing the abbey ruins, its roots contained within a 1m (3ft) high circular wall: the 'Omega' of Ogham hinting at what lies inside...

'What are the roots that clutch,
what branches grow

Out of this stony rubbish? Son of man,

You cannot say, or guess,
for you know only

A heap of broken images,
where the sun beats,

And the dead tree gives no shelter,
the cricket no relief,

And the dry stone no sound of water.'

From *The Waste Land*, T.S. Eliot, 1922

40

The Seven Oaks of Sevenoaks

Oaks (various), Sevenoaks, Kent, England

This is a tricky one. Although at various times there have indeed been seven oaks planted in Sevenoaks, all of those referred to in records seem to be predated by the town. As recently as 1902, seven oaks were planted around the Vine cricket ground (which dates back to 1773 and is the first place in England where a game of cricket was played with three stumps instead of two). These seem to be replacements – contemporary paintings and photography from 1900 depict 'The Seven Oaks' at the same location. After the Great Storm of 1987, all but one of the 1902 trees were uprooted. The six felled trees were replaced, but after vandalism and more replanting, there are now – rather confusingly – nine oaks on the site.

Local lore states that the town was named after a Saxon chapel called Seouenaca in the grounds of Knole Park. It is possible that this name is in fact derived from the fact that seven oaks stood near the chapel in about AD 800. (The Middle English for seven is recorded both as *seofan* and *seouen*.) However, it wasn't until the thirteenth century that a market town is recorded as existing around the Knole estate.

And then there's Sir William Sevenoak, or Sevenoke, who became Lord Mayor of London in 1418, and whose life followed almost as many twists and turns as the man who replaced him as Lord Mayor the following year – Dick Whittington. Legend has it that Sevenoak was discovered by Sir William Rumschedde as an orphan beneath a group of seven oak trees near the small hamlet of Riverhill in Kent. After serving as an apprentice grocer in London, he became first Sheriff and then Lord Mayor of London, and amassed considerable personal wealth, which he used to fund twenty almshouses as well as a free school in 1432 'at Sevenoaks' (originally Queen Elizabeth's Grammar School, now Sevenoaks School). Although the school is now far from free, Sir William's charity still lives on in the form of a scholarship scheme.

So, to recap: there are plenty of oaks in Sevenoaks. At various times since AD 800 there have been seven oaks in a group, in a number of locations from Knole Park to Riverhill and the Vine. Which trees the town is named after is uncertain – but it's unlikely, sadly, that any are still standing.

41

Shakespeare's Mulberry Trees

Mulberry, Stratford-Upon-Avon/Hastings, England

'Shakespare Mulberry Tree

In the Garden, at the back of East Cliff House, is a large Mulberry Tree, covering a space of 25 feet square, (said to be a descendant of Shakespeare's Mulberry Tree, at Stratford-upon-avon,) and planted by the great Actor Garrick, when he was on a visit to his friend Mr. Capel.

It is a curious fact, that there is at this time, a Cup made from the Mulberry Tree, planted by Shakespeare, in the possession of a lady at Hastings, which was a present to her Mother from Garrick in 1776; and also a Cup, turned this year from the Mulberry Tree above mentioned, planted by Garrick, in the possession of J.G. Shorter, Esq. [sic]'

Hastings and St Leonards Guide
(P.M. Powell, Eighth Edition, 1830)

It's hard to miss the site of New Place, the Stratford-upon-Avon home that playwright William Shakespeare bought in 1597: just follow the coachloads of tourists traipsing towards it with their selfie sticks. The tourists are nothing new. In 1756, some 143 years after the great bard's death, the steady stream of visitors to the house prompted the then owner of New Place, the Reverend Francis Gastrell, to take an axe to the magnificent black mulberry tree that stood in its garden, tired of the constant requests to see it. The locals were incensed. James Boswell, the great eighteenth-century diarist and biographer, recoiled at Gastrell's 'gothick barbarity'.

The reverend later applied for permission to extend his garden, but not only was this application rejected by the local authorities, they also increased his taxes. This was the final straw for Gastrell: in 1759 he went one step further, and had Shakespeare's timber and brickwork house – all twenty rooms and ten fireplaces – raised to the ground. He promptly left town (pursued by angry mob, stage left), but thankfully the story of Shakespeare's mulberry tree did not end with him.

It's worth noting at this point that at the turn of the seventeenth century there was a vogue for black mulberries, which had become something of a status symbol following a drive by King James I (he of Bible fame). Since silkworms will only eat mulberry leaves, the king saw an opportunity to kick-start an English silk industry to rival that of France, by encouraging wealthy landowners to plant mulberries. As Shakespeare didn't personally take ownership of New Place until 1602, this tallies with the theory that it was he who planted the famous mulberry in his garden. Indeed, several of his plays refer to the mulberry, even hinting at his horticultural expertise by noting particular details such as the staining qualities of its berries ('Now humble as the ripest mulberry/That will not hold the handling.' *Coriolanus*), Act III, Scene II.

Sadly, the king's plans came to naught – chiefly because silkworms prefer to feed on the leaves of white mulberries rather than the black variety, of which he imported 10,000 saplings. Some say that French arborists deliberately misled him in this detail – we couldn't possibly comment. He could at least claim the satisfaction of lending his name to his own species – *Morus nigra* 'King James' – that originated from Chelsea Physic Garden and also filled a 1.6-ha (4-acre) orchard close to the present site of Buckingham Palace.

As for Shakespeare's mulberry, as mentioned above, its story doesn't end with Gastrell's axe. Thanks to the entrepreneurial spirit of several local businessmen – chiefly Thomas Sharpe, as well as John Marshall and George Cooper – the wood of the felled tree was bought and used to fashion a range of souvenirs, which varied from high-value items such as tables and chairs, to more affordable objects including snuff boxes and tea caddies (still certainly better than a T-shirt and a key ring). Unfortunately for Sharpe, Marshall, Cooper, *et al*, the number of products on the market (many marketed as *memento mori* – a clever pun on *morus*, the Latin name for mulberry) – meant customers started to grow suspicious of their authenticity. To counter the claims of fakery, Sharpe took to adding a note of provenance to every item he sold. A message in one particular box from Sharpe, identified by the Shakespeare Birthplace Trust, takes pains to point out: 'This box was made of the real mulberry tree planted by Shakespeare in Stratford upon Avon just after it was cut down and before it was used up at the time of the Jubilee, when much fictitious mulberry wood supplied its place, for the purpose of memorial articles.'

One of the more reputable items, however, was a table from antiquarian furniture dealers H. Blairman & Sons, made in 1824 but believed to incorporate wood from Shakespeare's mulberry – which is inlaid with the initials 'WSMT' (for 'William Shakespeare's Mulberry Tree') as well as the date 1609, the year the tree was planted. Significantly, the provenance for this piece notes that it was once owned by the great (arguably the greatest) Shakespearean actor, David Garrick (1717–79).

Garrick's part in the story is significant. He would have been thirty-nine in 1756, the year that Shakespeare's mulberry was felled – and already hailed as the greatest actor of his (or any) age. As such, he was no stranger to Stratford-upon-Avon. The 'jubilee' mentioned in Sharpe's note above refers to the three-day celebration that Garrick organized in 1769 (presumably marking ten years since New Place was destroyed – unless it was arbitrarily picked to represent either the 205th anniversary of Shakespeare's birth or the 153rd anniversary of his death). During this jubilee, a cutting from Shakespeare's original mulberry was planted in New Place Garden, which still stands today (follow the selfie sticks). Other mulberries throughout the town also claim to have originated from cuttings of the Shakespeare tree – and there seems little reason to doubt these assertions, if only because the original would undoubtedly have been one of the first of its kind in the country.

Shakespeare's New Place was reopened to the public in 2016 (a rather more neat 400 years after his death), complete with a bronze sculpture of his original tree, a semi-circle of tourist-friendly benches protected by an arc of pleached trees, and accessed via an oak and bronze doorway that marks the original threshold of the house.

But there are other, living monuments to the great playwright's life. As recorded by the *Hastings and St Leonard's Guide* of 1830, Garrick's cutting travelled further afield than the outskirts of Stratford. Another tree seems to have made its way to the garden of Edward Capel (1713–81), the pre-eminent Shakespearean critic of his age. It would take another book to explain how Capel's tireless life's work was focused on correcting the errors that had crept into contemporary versions of Shakespeare's canon in the century since he passed away. Suffice to say, Garrick greatly admired him, and as a symbol of his affection, brought a cutting from the original tree to Capel's house on the slopes of East Cliff in Hastings.

The original – or perhaps its progeny – can still be found in the windswept Hastings landscape, accessible via a funicular from the foot of East Hill, near the fisherman's huts on Rock-a-Nore. Notably, another black mulberry, believed to be itself a cutting from the Garrick cutting, grows in the garden of a house on the opposite slope in Hastings Old Town, now owned – appropriately enough – by Craig Sams, former chairman of the Soil Association.

Stevenson's Yew

Yew, Colinton Church Manse,
Edinburgh, Scotland

'The river, on from mill to mill,

Flows past our childhood's garden still;

But ah! we children never more

Shall watch it from the water-door!

Below the yew – it still is there –

Our phantom voices haunt the air

As we were still at play,

And I can hear them call and say:

"How far is it to Babylon?"

To Minnie, Robert Louis Stevenson

This yew is still there, over 130 years since the great Scottish author Robert Louis Stevenson immortalized it (the lines above first appearing in his collection *Penny Whistles* in 1885). Before going on to write such classics as *Treasure Island* and *The Strange Case of Dr Jekyll and Mr Hyde*, Stevenson would spend his holidays at his grandfather's manse (the name given to a Presbyterian minister's home) in the Colinton suburb of Edinburgh.

Stevenson was a sickly child and was often privately tutored due to his absences from school. Surprisingly, he didn't learn to read until the age of seven or eight, but he was a keen storyteller – a trait that was encouraged by his father, Thomas, a frustrated author himself, who paid for the printing of his son's first publication, *The Pentland Rising: A Page of History, 1666*, which the young Robert wrote at the age of sixteen, in 1866. The subject was the Covenanters' rebellion of that year (which, incidentally, is closely linked to another tree mentioned elsewhere in this book: the Covenanter's Oak in Motherwell – see pp.56–57).

One obvious explanation for Stevenson's precocious interest in religion would be his maternal grandfather, Lewis Balfour, Presbyterian minister for the Parish of Colinton. Indeed, he even mused in his 1887 biography *Memories and Portraits*, 'Now I often wonder what I inherited from this old minister. I must suppose, indeed, that he was fond of preaching sermons, and so am I, though I never heard it maintained that either of us loved to hear them.'

Given this reticence, perhaps it is more likely that his childhood nurse, Alison 'Cummy' Cunningham, was the greater influence: a devout Calvinist, she would read him tales of the Covenanters alongside other religious texts as he lay ill.

But back to the yew tree. Despite his recurring illnesses, the young Stevenson was as adventurous as any young boy, and recalls climbing in – and swinging from – the boughs of the yew in his grandfather's garden. Today, a modern swing is attached to the tree, but there are marks revealing where an older swing was once fixed – most likely the one Stevenson once played on.

The sickly child grew into a gaunt but brilliant man. He was always dogged by respiratory trouble (modern diagnoses suggest he suffered from bronchial problems – possibly caused by an abnormal enlargement of the lung, or from sarcoidosis, a disease that affects the lungs and can cause shortness of breath, wheezing and coughing. Stevenson died in 1894, at the early age of forty-four, probably from a cerebral haemorrhage after collapsing while opening a bottle of wine. (His last words – to his wife – were: 'Does my face look strange?')

One hopes that in his final hours he was no longer troubled by the religious nightmares that plagued his childhood fevers – fed no doubt by Nurse Cummy. But perhaps his thoughts turned once more to the yew that gave him so much joy and solace. Amongst his surviving personal letters is one addressed to his close friend Fanny Sitwell from June 1874:

> '*Night*. – I suppose I must have been more affected than I thought; at least I found I could not work this morning and had to go out. The whole garden was filled with a high westerly wind, coming straight out of the hills and richly scented with furze – or whins, as we would say. The trees were all in a tempest and roared like a heavy surf; the paths

all strewn with fallen apple-blossom and leaves. I got a quiet seat behind a yew and went away into a meditation. [...] whenever there came a blink of sunshine or a bird whistled higher than usual, or a little powder of white apple-blossom came over the hedge and settled about me in the grass, I had the gladdest little flutter at my heart and stretched myself for very voluptuousness.'

43

The Suffrage Oak

Oak, Glasgow, Scotland

The Representation of the People Act, 1918, introduced a major reform to the electoral system of Great Britain and Ireland. In effect, it extended the right to vote to all men over the age of twenty-one (whether or not they owned property) and, for the first time, to women over thirty. Despite the long – and increasingly militant – struggle to establish 'Votes for Women' in Britain, the act still carried a number of caveats, and it would be another decade before the Representation of the People (Equal Franchise) Act, 1928, afforded electoral equality to all citizens over the age of twenty-one, regardless of their gender.

Several monuments are dedicated to the leaders of the movement for women's suffrage (which, despite its similarity to the word 'suffer', simply means the right to vote in public elections). But, among the statues and plaques, one is notable for its stoic elegance. Standing proudly on Kelvin Way in Glasgow (not far from Glasgow University) is the Suffrage Oak – a tree planted on 20 April 1918 to mark the passing of the act into law. A tribute to the countless women who marched on the streets of Glasgow, at great personal and reputational risk, it stands as a reminder of what was achieved by the 'suffragettes' (a term originally used disdainfully by the journalist Charles E. Hands in the *Daily Mail*, which was swiftly adopted by the campaigners).

Increasingly frustrated by the stalling and inaction of MPs, the movement adopted a highly visible campaign of direct action. Emmeline Pankhurst (1858–1928) founded the Women's Social and Political Union (WSPU) in 1903, later recalling their desire 'to be satisfied with nothing but action on our question. "Deeds, not words" was to be our permanent motto...'

As tensions rose, a secretive group of WSPU members known as 'The Bodyguard', was set up to protect Pankhurst and other prominent suffragettes – both from physical attack and arrest. One of their more significant actions was the 'Battle of Glasgow' (1914), in which some thirty members physically brawled with fifty policemen in front of 4,000 onlookers at St Andrew's Hall in Glasgow.

The memory of these brave women lives on in the Suffrage Oak, which was voted Scotland's Tree of the Year 2015 – a notable and appropriate case in which the tree's age and appearance was considered less worthy of praise than what it represents.

A plaque placed by Glasgow District Council in 1995 reads simply: 'This oak tree was planted by Women's Suffrage Organisations in Glasgow on 20 April 1918 to commemorate the granting of votes to women'.

In October 2017, the tree was damaged by the effects of Storm Ophelia and its height and canopy had to be reduced by the city council for safety reasons. Far from being diminished, however, the removed sections of branch were given new life – being gifted to Glasgow Women's Library to be fashioned into items to be sold in memory of the women who fought – often literally – for equality.

44

The Tolkien Trees

Beech, Avebury, Wiltshire, England

The village of Avebury in Wiltshire is encircled by a series of ancient standing stones, laid out in three concentric circles on earthworks and thought to date to as early as 2850 BC, making them about the same age as the stone circle at nearby Stonehenge.

Close to the eastern entrance of the outer bank, on slightly raised ground, are four magnificent copper beech trees, as if standing guard over the sacred site. The four trees form a natural enclosure, their canopies joining to form a foliage roof and their knotted root masses creating a textured carpet on the earth floor. Their branches are adorned by colourful prayer ribbons, with more bejewelled offerings, markings and inscriptions embedded in their trunks by those who have passed under them.

These trees are said to be the inspiration for the Ents, a race of treelike beings in Tolkien's *The Lord of the Rings* trilogy, who act as benevolent shepherds of the forest. Their name is derived from *ettin* or *eoten*, the Anglo-Saxon word for giant, and Tolkien also drew from various myths of talking trees in folklore – particularly the oracular trees that Druids conversed with and consulted.

While Tolkien never overtly referred to the Avebury trees, he was known to have visited them and is said to have sat under their boughs, sketching. In *The Lord of the Rings*, he describes Treebeard (the *de facto* leader of the Ents, and oldest living creature in Middle Earth) as resembling 'an oak or beech'. In a letter to W.H. Auden, he says of the Ents, 'Their part in the story is due, I think, to my bitter disappointment and disgust from schooldays with the shabby use made of Shakespeare of the coming of "Great Birnam wood to high Dunsinane Hill."' (See pp.40–41.)

Clearly, Tolkien's fierce imagination was let down by the rather prosaic method by which the trees of Birnam wood made their way up the hill – far better that they should uproot and stride across the

'I may love him, I may love him, for he is a man, and I am only a beech-tree.'

From *Phantastes*, George MacDonald

countryside. In this, Tolkien also admitted the influence of the moving, talking trees in George MacDonald's *Phantastes*.

The ability of the Ents to uproot and walk amongst their allies is influenced by legends of willows stalking human travellers – most likely because of their silhouette, which resembles that of a crouching person.

Tolkien details how, once they reach an advanced age, Ents would become 'treeish', laying down roots and losing their articulation to become a wizened, static part of nature – just like the beeches at Avebury.

45

The Tolpuddle Martyrs' Tree

Sycamore, Tolpuddle, Dorset, England

Britain's trade unions are typically associated with its industrialized cities, but the roots of an organized labour movement can be traced to rural workers, squeezed out by the Inclosure Acts of the late eighteenth and early nineteenth centuries. These acts essentially handed areas of common land to landowners by creating legal property rights to areas that were formerly under collective control. In simple terms, these acts effectively denied the traditional right to graze and re-established a feudalist system of land ownership.

In all, 5,200 Inclosure Acts were passed over a period of 300 years (from 1604–1914), with 2.8 million ha (almost 7 million acres) of previously open land being enclosed. (The word 'inclosure' is simply an early spelling of 'enclosure'.) But in this instance, we are only concerned with a tiny patch of common ground – one of the smallest of all the properties now maintained by the National Trust – and, to be more specific, one tree: a sycamore in the Dorset village of Tolpuddle, near Dorchester, thought to date from 1680.

In 1834, The Friendly Society of Agricultural Labourers was formed by six men from Tolpuddle, led by the thirty-seven-year-old Methodist preacher George Loveless. The Industrial Revolution had intensified the flurry of Inclosure Acts, with a series of consolidation laws being passed in order to create a more efficient use of land and labour in response to the rapidly-changing working landscape.

The six men – Loveless, his brother James, his brother-in-law Thomas Standfield, Thomas's son John, James Brine and James Hammett – met beside the sycamore tree in the village square and drew up plans 'to preserve ourselves, our wives and our children from utter degradation and starvation'. Despite the average weekly family expenditure of the time being over thirteen shillings, the men had already seen their wage bill steadily fall from nine shillings to seven, with rumours that it would soon fall to six: less than half the basic subsistence income. In short, the new laws meant that these agricultural workers could no longer provide for their families. In response to this perceived exploitation of their labour, the group avowed they would refuse to work for less than ten shillings a day.

At the time, the government had been rocked by a succession of Swing Riots, in which agricultural labourers across the south of England had destroyed farm machinery and maimed cattle in protest against threats to their livelihood. Landowners – and the government – were looking for the excuse to set an example. And the Tolpuddle Six, it was clear, weren't prepared to accept their lot. They sought the support first of the Vicar of Tolpuddle, then of Robert Owen, leader of the Grand National Consolidated Trades Union.

In an official complaint about the group to the Whig government, local landowner James Frampton cited the obscure 1797 Incitement to Mutiny Act. Despite the fact that the act was aimed specifically at 'Persons serving in His Majesty's Forces by Sea or Land' as a direct response to the Spithead and Nore mutinies (that took place on board Royal Navy ships during the war with France in that year), it was deemed applicable to six farm labourers in Dorset.

Found guilty of entering into a 'sworn union' (the phrase taken directly from the Mutiny Act), the six were sentenced to seven years' transportation to Australia. Delivered amidst a groundswell of opinion in favour of trade unionism, the sentence was perceived to be unduly harsh: indeed, the forming of unions was no longer illegal in 1834. In response, thousands of people marched in London, while a petition signed by 800,000 people was delivered to Parliament.

The words of George Loveless, written in prison while awaiting his sentence to be carried out, became a rallying cry for Britain's new movement of politicized working classes: 'We raise the watchword: liberty. We will, we will, we will be free!'

Within three years, all six of the Tolpuddle Martyrs had been pardoned and released. On his return, Loveless wrote an account of their case, *The Victims of Whiggery*, while Thomas and John Standfield,

James Loveless and Brine co-wrote *The Horrors of Transportation*. Both publications are now seen as essential texts in the birth of trade unionism. By 1867, a Royal Commission decreed that the establishment of trade union organizations was of advantage to both employees and – crucially – employers. In 1868, the Trades Union Congress was founded. Three years later, in 1871, the Trade Union Movement was legalized, inspired by the words of the great philosopher and political economist John Stuart Mill: 'If it were possible for the working classes, by combining among themselves, to raise or keep up the general rate of wages, it needs hardly be said that this would be a thing not to be punished, but to be welcomed and rejoiced at.'

Almost 200 years after six Dorset farm workers met on a village green to establish that principle, it is still seen as a bedrock of modern liberal democracy. The sycamore tree under which those men stood may be slightly stunted by age in the intervening years, but it still stands proud, conserved by the National Trust for its historical significance.

'That each day I may walk unceasingly on the banks of my water, that my soul may repose on the branches of the trees which I planted, that I may refresh myself under the shadow of my sycamore.'

Egyptian tomb inscription, circa 1400BC

The Trafalgar Woods and the Nile Clumps

Various species throughout the UK/Beech,
Wiltshire, England

Achille	*Dreadnought*	*Polyphemus*
Africa	*Entreprenante*	*Prince*
Agamemnon	*Euryalus*	*Revenge*
Ajax	*Leviathan*	*Royal Sovereign*
Belleisle	*Mars*	*Sirius*
Bellerophon	*Minotaur*	*Spartiate*
Britannia	*Naiad*	*Swiftsure*
Colossus	*Neptune*	*Temeraire*
Conqueror	*Orion*	*Thunderer*
Defence	*Phoebe*	*Tonnant*
Defiance	*Pickle*	*Victory*

The final name in this list should prove the giveaway even if you weren't paying attention in history lessons: these are the names of the twenty-seven ships of the line (plus six support ships) of Nelson's fleet in the Battle of Trafalgar. It also happens to contain the names of twenty-seven woods planted across Great Britain and Northern Ireland by the Woodland Trust as part of its Trafalgar Woods Project. Started in 2005 to mark the bicentenary of Nelson's famous victory over the Spanish and French fleets, the project has seen over 250,000 new trees planted, celebrating the link between British timber and its maritime history. The landmark Victory Wood, named after Lord Nelson's 104-gun flagship, was planted in Kent overlooking the Isle of Sheppey, near the mouth of the Medway, close to the site where HMS *Victory* was constructed.

It is estimated that 6,000 trees went into the building of HMS *Victory*, 90 per cent of which were oak, with pine, fir, elm and lignum vitae used in its finishings. Construction began in 1759 but, thanks in part to delays due to the end of the Seven Years' War, the frame of the *Victory* was uniquely left to season for three years (rather than the standard few months), which many believe was a key factor in the ship's longevity: at the time of the Battle of Trafalgar, she was already forty years old, having been officially launched on 7 May 1765. In fact, the *Victory* was serving as a hospital ship (as she had been ruled unfit for service as a warship) when she was extensively refitted in 1800 to make up for the loss of the HMS *Impregnable* off Chichester.

The Trafalgar Woods aren't the first time that trees have been planted to commemorate one of Nelson's victories: the Nile Clumps on Salisbury Plain mark his 1798 defeat of the French in the Battle of the Nile. Planted by Charles Douglas, 6th Marquess of Queensberry, on his own estate near Stonehenge, the beech trees were built in clumps representing the positions of the French and British ships in the battle. The original clumps are now sadly depleted from a combination of natural wastage and the building of the A303 (see pp. 172–173): of an original twenty-six clumps, seventeen still remain.

The trees are sometimes falsely referred to as the Trafalgar Clumps, although since they are planted in two parallel lines, it is obvious that they do not refer to the position of the ships at Trafalgar.

Nelson's unorthodox tactic at Trafalgar was to split his ships into two lines and drive them through the Franco-Spanish fleet, splitting the long enemy formation at two points. Rather than coming alongside the enemy in single file to engage them, as was the norm, Nelson's revolutionary plan was designed to create maximum confusion. Contemporary reports suggest that many of his peers thought it was a rash and even primitive approach, although Nelson himself reports that after describing the plan to his captains, 'it was like an electric shock. Some shed tears, all approved – "It was new – it was singular – it was simple!"'

Or so the myth goes. In his extraordinary book *Seize the Fire: Heroism, Duty and the Battle of Trafalgar*, Adam Nicolson states that, 'The reason for the British success at Trafalgar was not tactical. The tactics were immensely weak. The success depended on the independent ferocity and fighting aggression of each British ship and on the example of leadership given by Nelson to his captains.'

Nicolson cites a letter from Nelson to Keats in explaining how 'confusion and chaos was... Nelson's chosen method of battle':

> 'I think it will surprise and confound the Enemy. They won't know what I am about. It will bring forward a pell-mell Battle, and that is what I want.'

So if the trees on Salisbury Plain had been planted in honour of the later battle, they would more closely resemble a 'hash' symbol than two roughly parallel lines. And even then, the muddle of interlocking trees would be so chaotic as to barely make any pattern at all. Perhaps a wild wood, planted erratically and left to run wild, with its mixture of species and its natural selection of survival, is a more fitting tribute after all.

47

The 'Trees To Remain'

Various species: Hampshire, Wiltshire,
Dorset, Somerset, Devon, England

The A303 is one of the most brilliantly eccentric roads in Britain. Starting just outside Basingstoke in Hampshire, it flirts around the edges of Dorset and Wiltshire before petering out near the Somerset/Devon border at Honiton, constantly denying opportunities to overtake as it meanders towards the sunset in the west.

It's not a road to take if you're in a rush. Bottlenecks abound as it swells and contracts between dual carriageway and single lane, with confused lorry drivers following their sat-navs into ever-smaller lanes. Occasionally, as it rises and dips over Salisbury Plain, low-lying fog can suddenly envelop you like an eerie Arthurian mist.

It is, of course, most famous for being the road that passes Stonehenge, prompting a constant tailback as rubber-necking drivers slow down to view a landmark that has drawn visitors for millennia. (As the artist Jeremy Deller helpfully pointed out in *Wiltshire Before Christ*, his 2019 joint exhibition with photographer David Sims and fashion label Aries Arise, the A303 was 'built by immigrants' – a wry nod to DNA research from University College London that found many of the builders of Stonehenge originated from modern-day Greece and Turkey.) But this road to ruin has countless other esoteric secrets to discover, as Tom Fort's charmingly curious book and BBC documentary *A303: Highway to the Sun* reveals.

Alongside the tales of medieval murder, Neolithic travellers and modern-day Druids, Fort unearths the original hand-drawn plans for the 'Andover bypass', dating from 1969, held at the Hampshire Record Office. Here, with their canopies drawn from above with painstaking attention to detail, are countless trees identified for preservation, each clearly marked 'Tree to Remain'.

It's heartening to know that, in planning offices across the area, as they plotted to bisect Harewood Forest, one of England's ancient woodlands in the lost county of Wessex, there was consideration for the history and heritage of one of Britain's oldest trackways – the Harrow Way, which has carried people from Dover in Kent to Seaton in Devon since the Stone Age.

Along the route of the A303 one also encounters arguably the country's oldest crossroads (between Harrow Way and the Great Ridgeway) as well as segments of the famous Roman Fosse Way. But where the Fosse Way describes an almost perfect straight line diagonally across England (from Exeter to Lincoln), the A303 deviates with the landscape, wherever possible respecting – and even dictated by – the topography.

The 'Trees To Remain' are a prime example – nestling in grassy central reservations, proudly denoting bends in the road, still marking their territory as modern traffic flows either side of them. These ancient trees and tiny remnants of England's wildest woodland have witnessed man's technological advances – from steam-powered carriages to railways to motor cars – but somehow they have endured, defiant against the march of progress.

48

The Wesley Beeches

Beeches, Lambeg, County Down, Northern Ireland

Theologians might often disagree on the origins of various religions. It's often impossible to determine the exact point at which a new religious movement is first founded: questions of faith tend to evolve over time, their veracity debated and dissected. However, we do know exactly when Methodism was founded, at 8.45pm on 24 May 1738.

That, according to his own account, is the precise moment when John Wesley had what he called his 'Aldersgate experience'. Wesley had just returned from what he deemed to be a failed venture to the American colonies to re-establish 'primitive Christianity' as minister to the new parish of Savannah, Georgia. In a depressed state of mind, he attended a meeting of Moravian Protestants in Aldersgate Street, London. During a reading of one of Martin Luther's Epistles, Wesley describes having a moment of epiphany. 'I felt my heart strangely warmed,' he later wrote. 'I felt I did trust in Christ, Christ alone for salvation, and an assurance was given me that he had taken away my sins, even mine, and saved me from the law of sin and death.'

Aldersgate Day is still celebrated every 24 May by the Methodist church, marking the beginnings of Wesley's evangelical revivalist movement.

Wesley's sermons, largely delivered outdoors, emphasized sanctification, rejecting predeterminism and teaching instead that salvation is available to all (attained through living according to certain moral principles and carrying out acts of charity). Methodists were often pioneers of social change, championing the abolition of slavery, prison reform and supporting female preachers. Wesley's focus on missionary work meant that Methodism spread widely across the British Empire as well America, and today there are an estimated 80 million followers.

Despite initially being barred from preaching in many parishes (hence the tendency for outdoor meetings), Wesley always maintained that Methodism lay firmly within the traditions of the Anglican Church. During a trip to Northern Ireland, he demonstrated this belief by means of physical metaphor. On 10 June 1787, while preaching on Chrome Hill near Lisburn in County Down, he twisted together two beech saplings, using them to symbolize how Methodism and the Anglican Church of Ireland were intertwined.

In the 200 years since, the two fully-grown beech trees are now interlocked, their gnarled branches wrapped together in an inseparable arch. Developing the metaphor further, the trees have even been strengthened by their union, which has added greater stability against the elements, overcoming a vulnerability usually common to beeches due to their shallow roots. They also come to leaf earlier than surrounding trees, drawing further attention to Wesley's dramatic creation.

49

The W.G. Grace Tree

Oak, Sheffield Park, East Sussex, England

Situated near Uckfield in East Sussex, Sheffield Park hosts one of the earliest cricket grounds of the modern era. It was built in 1845 – six years after the founding of Sussex, the first county club in the country (and a year before the first recorded international cricket match, which took place – improbably – between the USA and Canada at St George's Cricket Club, New York).

The first match played at Sheffield Park took place between the local villages of Fletching and Chailey. Opening the batting for Fletching was Henry Holroyd, aka Viscount Pevensey (later Lord Sheffield), who was at the time a thirteen-year-old schoolboy on holiday from Eton. Upon inheriting the Sheffield Park estate in 1876, he oversaw a complete reconstruction of the ground – said at the time to be one of the most beautiful in England.

Sheffield Park holds several claims to fame. It staged the first ever matches in England involving teams from India (in 1886) and South Africa (in 1894). The first of five Australian visits to play the Lord Sheffield's XI was in 1884 and, having built a rapport with the tourists, in 1891–92 Lord Sheffield took an England touring team to Australia at his own expense. To commemorate the trip, he donated the princely sum of £150 to the New South Wales Cricket Association to purchase a trophy and establish a competition between the leading states, to be named in his honour. To this day, the Sheffield Shield remains the main domestic cricket competition in Australia.

In 1896, a match against the Australians at Sheffield Park, featuring the Prince of Wales and the great W.G. Grace amongst the Lord's team, drew a crowd of some 25,000 people. It was during this match that Australian pace bowler Ernie Jones delivered a short-pitched 'bouncer' that met W.G. Grace so close to his face that witnesses – including fellow English players C.B. Fry and Stanley Jackson (who

'Druid and devilish deity and lean wild beast, harmless now, are revolving many memories with me under the strange, sudden red light in the old wood, and not more remote is the league-deep emerald sea cave from the storm above than I am from the world.'

From *The Heart of England*, Edward Thomas

was batting with Grace at the time) – remarked that the ball appeared to part the great man's beard. Indignant, Grace complained to Harry Trott, the Australian captain, asking, 'Here, what is all this?' Having been admonished by his captain ('Steady, Jonah!'), Jones turned to Grace and said, 'Sorry, doctor: she slipped!'

Whether the incident was a genuine accident or an attempt to 'rough up' the batsman – a good thirty-six years before England's Bodyline short-ball tactic that so incensed Australia – remains to be seen. Certainly, the fact that the story has been repeated down the years by players suggests a hint of sarcasm in Jones' apology. An early example, perhaps, of the 'sledging' that the great Australian teams of the 1990s became masters of.

Despite this incident, the matches against the Australians were said to have been played in a spirit of 'cordial happiness' – even when, on one occasion in 1890, Lord Sheffield's XI were bowled out for just twenty-seven runs (Grace accounting for twenty of those).

But an earlier game on the same ground, between a team led by Grace against Lord Sheffield's XI, explains the existence of a tree that bears the great cricketer's name. The historic oak, distinctive due to the iron girdle that circles its trunk, has been known as 'W.G. Grace's tree' since he struck it 'full toss' with an epic six during a match there in July 1883. This match is still recorded as being the highest-ever team total in a first-class game at the time – although a moderate one by today's standards – of 266.

Sheffield died in 1909 – unmarried and with no heirs – and the estate was sold; its cricket pitch and magnificent gardens slowly fell into disrepair. During the First World War, the land was requisitioned for farming and the pitch was ploughed up. Further damage was sustained during the Second World War when Sheffield Park was used as the headquarters of a Canadian armoured division.

Eventually bought by the National Trust in 1953, the ground has recently been restored to its former glory and is now home to the Armadillos Cricket Club.

In 2009, to mark the centenary of Sheffield's death, a commemorative match was played between an Old England XI and Lord Sheffield's Australian XI – who brought with them the original Sheffield Shield.

No trees were hit and no beards were parted.

The Wilberforce Oak

Oak, Hayes, Kent, England

On 12 May 1787, William Pitt the Younger, the Prime Minister of Great Britain (and, by 1801, of the new United Kingdom) met with William Wilberforce, the independent Member of Parliament for Yorkshire, on Keston Common, part of the estate of Holwood House, Pitt's Decimus Burton-designed home, near Hayes in Kent.

Here, under an oak tree, the two stopped, deep in discussion. Wilberforce recalls the exchange in an entry in his personal diary:

'At length, I well remember after a conversation with Mr. Pitt in the open air at the root of an old tree at Holwood, just above the steep descent into the vale of Keston, I resolved to give notice on a fit occasion in the House of Commons of my intention to bring forward the abolition of the slave-trade.'

These remarks are now recorded for prosperity in the inscription in a stone bench, erected in 1862 – and known as the Wilberforce Seat – that sits at the stump of the original oak, long since gone. A new sapling has been planted in place of the Wilberforce Oak, a slightly puny marker of one of the prouder moments in Parliamentary history: the spot at which the seeds of the Slave Trade Act of 1807 – and subsequently the Slavery Abolition Act of 1833 – were sown.

Already four years into his role as Britain's youngest Prime Minister (a position which he assumed in 1783 at the age of just twenty-four, and held for a total of nineteen years over two periods of office), Pitt's reputation was one of a reformist, amidst great geopolitical pressures – not least the Napoleonic Wars. Wilberforce was effusive in his praise for Pitt. 'For personal purity, disinterestedness and love of this country, I have never known his equal', he stated.

Much the same could be said of Wilberforce. Like Thomas Clarkson, the abolitionist who had first persuaded him to join the cause in 1787 (the same year as the meeting under the oak),

Wilberforce was inspired by his faith. A recent convert to evangelicalism at the time, renouncing his earlier life of gambling and drinking, he saw slavery as an affront to God and humanity.

In a speech before the House of Commons on 18 April 1791, Wilberforce passionately laid out his case:

> 'Never, never will we desist till we have wiped away this scandal from the Christian name, released ourselves from the load of guilt, under which we at present labour, and extinguished every trace of this bloody traffic, of which our posterity, looking back to the history of these enlightened times, will scarce believe that it has been suffered to exist so long a disgrace and dishonour to this country.'

His position was not a populist one. The catalyst for the anti-slavery movement was the Committee for the Abolition of the Slave Trade, founded by nine devout Quakers, plus three Anglicans (including Clarkson). Those MPs who took up the cause were largely derided, and sarcastically referred to as 'The Saints'. It took twenty years of campaigning for the Slave Trade Act to become law – and even then, its parameters were limited to prohibiting the slave trade within the British Empire. Technically it didn't declare slavery illegal, although abolitionists pointed to a judgement of 1772 (*Somerset v. Stewart*), which ruled that slavery had never been supported in English law. The case revolved around James Somerset, 'an enslaved African' who had been purchased by a British customs officer in Boston, in what was at the time part of the British Province of Massachusetts Bay. Having been brought back to England by Stewart, Somerset escaped in 1771.

When Stewart caught him and had him imprisoned, with orders that he be sold for labour on a Jamaican plantation, Somerset's three British godparents from his baptism in England brought his case before William Murray, 1st Earl of Mansfield and Chief Justice of the King's Bench.

Lord Mansfield was well aware of the significance of the case. In his summary, he announced, 'We feel the force of the inconveniences and consequences that will follow the decision of this question. Yet all of us are so clearly of one opinion upon the only question before us, that we think we ought to give judgment...'

That judgement being that Somerset should be 'discharged' and that chattel slavery (treating a person as the personal property of another) was 'unsupported by the common law of England and Wales'.

Crucially, however, this ruling – combined with the Slave Trade Act – still fell short of declaring slavery illegal across the British Empire. It took another quarter of a century until the Slavery Abolition Act was passed in 1833.

In May of that year, the Whig government introduced the Bill, formally saluting Wilberforce for his role in seeing it through. Sadly, however, Wilberforce didn't live to see it become law. On 26 July 1833, in failing health following a severe bout of influenza, Wilberforce learned that government concessions had guaranteed the bill would be passed. Three days later, on 29 July, he passed away: his life's work done – as the final words of his famous 1789 'Abolition Speech' to the House of Commons attest:

> 'I confess to you sir, so enormous so dreadful, so irremediable did its [slavery's] wickedness appear that my own mind was completely made up for the abolition. A trade founded in iniquity, and carried on as this was, must be abolished, let the policy be what it might – let the consequences be what they would, I from this time determined that I would never rest till I had effected its abolition.'

'Brian has a thing about trees. Can't stop looking at them. I've been abroad with him when he's been in rapture over an avenue of pine trees. "Look at them, Bill," he'd say. "Aren't they beautiful! People don't appreciate beauty these days. They look at everything but they don't really see. Who really looks at trees and sees their shapes and colours? They're magic! That's what it's all about!"'

Bill Clough, on his brother Brian, manager of the English football clubs Hartlepool United, Derby County, Brighton & Hove Albion, Leeds United (briefly) and Nottingham Forest

Index

An esoteric BIBLIOGRAPHY of books that helped to inspire this book, from childhood onwards...

Out of The Woods, Will Cohu (Short Books, 2007).
For being the B.S. Johnson of tree guides.

The Worm Forgives the Plough, John Stewart Collis (C. Knight, 1973).
His two classics – *While Following the Plough* (1946) and *Down to Earth* (1947) – published as one volume: for his unromanticized view of nature.

The English Landscape in Picture, Prose and Poetry, collected and arranged by Kathleen Conyngham Greene (Ivor Nicholson & Watson, Ltd. 1932).
For its countless inspiring quotations.

Nature, Ralph Waldo Emerson (James Munroe and Company, 1836).
From first tree to transcendentalism.

Severn & Somme, Ivor Gurney (Sidgwick & Jackson, Ltd, 1917).
For equating the exterior with the interior world.

War's Embers, Ivor Gurney (Sidgwick & Jackson, Ltd, 1919).
For making me see the woods of the West Country anew.

The Gardener's Labyrinth, Thomas Hill (1577; author's version Oxford University Press, 1987).
The first English book to attempt to rationalize the nation's peculiar mania for personal cultivation.

Meetings With Remarkable Trees, Thomas Pakenham (Orion Publishing Co., 1996).
For its categorization according to 'personality' over species.

The True Book About Trees, Richard St Barbe Baker (Frederick Muller Ltd, 1965).
For planting the idea of storytelling through trees.

The Heritage Trees of Britain & Northern Ireland, Jon Stokes and Donald Rodger (Constable, 2004).
A tick-box of trees to visit across the UK.

The Observer's Book of Trees, compiled by W.J. Stokoe (Frederick Warne & Co., 1937).
The archetypal idiot's guide, fits in the pockets of most reliable outerwear garments.

Man and the Natural World: Changing Attitudes in England 1500–1800, Keith Thomas (Allen Lane, 1983).
Particularly for how it tracks the change in attitude towards trees – from hostility to romance.

The Englishman's Country, edited by W.J. Turner (Collins, 1945).
For positioning the rural in the urban and the social in the natural.

Trees: A Ladybird Book, Brian Vesey-Fitzgerald (Ladybird Books, 1963).
If only for the silhouettes of British trees in winter in the end papers.

Trees of The British Isles in History & Legend, J.H. Wilks (Frederick Muller Ltd, 1972).
An invaluable and much-thumbed resource.

Orlando, Virginia Woolf (Hogarth Press, 1928).
Not least for its book-within-a-book, the epic poem *The Oak Tree* (or for being one of the greatest English novels in its own right).

Acknowledgements

Some words of thanks for helping in making this book become a reality.

The seed of an idea for this book was formed from countless walks in numerous woods, from my childhood onwards. It took root as a short-lived (and very primitive) blog over a decade ago, since when various plans to propagate it as a magazine column, email newsletter and social media account fell on stony ground.

I'm deeply indebted to Katie Cowan for her initial encouragement and enthusiasm in seeing how it could become a viable book – not to mention the work of Lucy Smith, Helen Lewis, Katie Hewett, Alice Kennedy-Owen, Isabelle Holton and many more of their colleagues at Pavilion for helping it to bear fruit. I think that's enough tree metaphors for now.

There are several people who saw the potential in the *Great British Tree Biography* from the start and who, in their various ways, helped to open doors, point me in the right direction, show support, or just keep my hopes up when I needed it. These include Richard Benson, the best editor I've ever worked for, a great writer and a lovely man; Jeff Barrett of Heavenly records and (more relevantly) *Caught by the River*, whose simple 'I'm in!' meant more than he could possibly have known; not forgetting Robin Turner and Steve Phillips, part of the wider CBTR orbit; Jude Rogers, whose opinion has been invaluable; Chris King for his knowledge, ideas and suggestions; and Sam Walton for helping me realise the only way is to write about what you love.

Thanks also to Keith, John, Steve and Craig – aka The Circle – for keeping me sane over all these years. Steve Wood, Simon Das and Kevin Braddock for their friendship and Marissa through it all.

A huge thank to Amy Grimes for her wonderful illustrations that lifted the whole project.

Finally, to my parents Jan and Alan for instilling in me a love of nature and for their constant support – and to Mr Fox for his early inspirations.

This book is dedicated to the memory of Jack and Iris Preston and wonderful walks among the trees with them – and to Tim Clark, whose laughter never made me feel more proud to be a writer.

First published in the United Kingdom in 2021 by Pavilion
43 Great Ormond Street
London
WC1N 3HZ

Illustrations by Amy Grimes

Seamus Heaney, 'The Wishing Tree', from *Opened Ground:
Poems 1966–1996* (2002). Copyright © Faber and Faber Ltd.

T. S. Eliot, *The Waste Land* (2019). Reproduced with
the kind permission of Faber and Faber Ltd on behalf of
The Estate of T. S. Eliot. Copyright © The Estate of T. S. Eliot.

ISBN 9781911641339

A CIP catalogue record for this book is available from
the British Library.

10 9 8 7 6 5 4 3 2 1

Reproduction by Rival Colour Ltd., UK
Printed and bound by 1010 Printing International Ltd., China

www.pavilionbooks.com

Publisher: Helen Lewis
Commissioning Editor: Lucy Smith
Editor: Izzy Holton
Design Manager: Alice Kennedy-Owen
Production Controller: Phil Brown